Nicotine Addiction in Britain

A report of the Tobacco Advisory Group
of The Royal College of Physicians

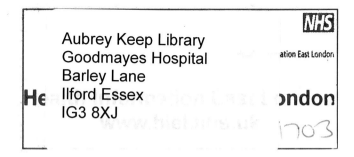

ROYAL COLLEGE OF PHYSICIANS OF LONDON

Membership of the Tobacco Advisory Group
of the Royal College of Physicians

John Britton (Chair)
Clive Bates
Kevin Channer
Linda Cuthbertson
Christine Godfrey
Martin Jarvis
Ann McNeill

Cover
photograph: Melanie Friend/Format
design: Merriton Sharp

Royal College of Physicians of London
11 St Andrews Place, London NW1 4LE

Registered Charity No. 210508

Copyright © 2000 Royal College of Physicians of London

ISBN 1 86016 1227

Typeset by Dan-Set Graphics, Telford, Shropshire
Printed in Great Britain by The Lavenham Press Ltd, Sudbury, Suffolk

Contributors

David Balfour *Reader in Pharmacology and Neuroscience, Ninewells Hospital, Dundee*

Clive Bates *Director, Action on Smoking and Health, London*

Neal Benowitz *Professor of Medicine, Psychiatry and Biopharmaceutical Sciences, University of California, San Francisco, USA*

Virginia Berridge *Professor of History, London School of Hygiene and Tropical Medicine*

John Britton *Professor of Respiratory Medicine, City Hospital, Nottingham*

Christine Callum *Statistician, Health Education Authority, London*

Kevin Channer *Consultant Cardiologist, Royal Hallamshire Hospital, Sheffield*

Linda Cuthbertson *Press and Public Relations Manager, Royal College of Physicians, London*

Jonathan Foulds *Senior Lecturer in Clinical Psychology, University of Surrey*

Christine Godfrey *Professor of Health Economics, University of York*

Peter Hajek *Professor of Psychology, St Bartholomew's and the Royal London Hospital*

Jack E Henningfield *Vice President, Research and Health Policy, Pinney Associates, Bethesda; Associate Professor of Behavioral Biology, Johns Hopkins University School of Medicine, Baltimore, USA*

John R Hughes *Professor of Psychiatry, University of Vermont, USA*

Martin Jarvis *Professor of Health Psychology, University College, London*

Ann McNeill *Strategic Research Adviser, Health Education Authority, London*

Lesley Owen *Senior Research Manager, Health Education Authority, London*

Martin Raw *Honorary Senior Lecturer in Public Health, Guy's, King's and St Thomas's School of Medicine, London*

Amanda Sandford *Research Manager, Action on Smoking and Health, London*

John Slade *Professor of Environmental and Community Medicine, University of Medicine and Dentistry of New Jersey, USA*

John Stapleton *Senior Lecturer, Institute of Psychiatry, London*

Ian Stolerman *Professor of Behavioural Pharmacology, Institute of Psychiatry, London*

Gay Sutherland *Honorary Consultant Clinical Psychologist, Institute of Psychiatry, London*

David Sweanor *Senior Legal Counsel, Smoking and Health Action Foundation, Ottawa, Canada*

Robert West *Professor of Psychology, St George's Hospital Medical School, London*

Sue Wonnacott *Reader in Neuroscience, University of Bath*

Foreword

In 1962, the Royal College of Physicians published its first report on the effects of smoking on health, drawing attention to the strong relationship between cigarette smoking and lung cancer. The report concluded that this association was probably causal, that smoking may also cause other diseases including chronic bronchitis and coronary heart disease, and that smokers may be addicted to nicotine.

In the years since that report was published, the true scale of the harm caused by smoking has become apparent. Smoking is now recognised as the single largest avoidable cause of premature death and disability in Britain and in most other economically developed countries, and probably the greatest avoidable threat to public health worldwide.

Public recognition of the health risks of smoking was probably one of the major factors underlying the progressive fall in smoking prevalence that occurred in Britain between the early 1960s and mid-1990s. However, recent data suggest that it is now beginning to stabilise in Britain at approximately one in four adults, whilst smoking in younger people is becoming more common. To achieve further marked reductions in smoking prevalence, it is therefore necessary to look in more detail at the factors that cause individuals to smoke, and to consider new methods of primary and secondary prevention.

This report addresses the fundamental role of nicotine addiction in smoking. It is now recognised that nicotine addiction is one of the major reasons why people continue to smoke cigarettes, and that cigarettes are in reality extremely effective and closely controlled nicotine delivery devices. Recognition of this central role of nicotine addiction is important because it has major implications for the way that smoking is managed by doctors and other health professionals, and for the way in which harmful nicotine delivery products such as cigarettes should be regulated and controlled in society. At a time when smoking still causes one in every five deaths in Britain, measures designed to achieve further reductions in smoking are clearly important and, if successful, will realise substantial public health benefits. It is time for nicotine addiction to become a major health priority in Britain. This report explains why.

February 2000
KGGM ALBERTI
President, Royal College of Physicians

Acknowledgements

The members of the Tobacco Advisory Group thank Lynn Koslowski for reviewing the manuscript, and Rachel Orme for editorial assistance.

The efficient and speedy production of this book was carried out by Amanda May, Suzanne Fuzzey and Diana Beaven of the RCP publications department, and the text was expertly copyedited by Dr Mary Firth. To all of them, our thanks.

Contents

KEY POINTS

- Smoking prevalence in Britain has declined during the past 50 years; this trend now appears to be stabilising

- In 1997 in Britain approximately one in four adults were cigarette smokers

- By age 15, one in four British children are regular smokers

- Smoking causes one in every five deaths in Britain, and the loss of more than 550,000 years of life before age 75

- The greatest impact of smoking on mortality is on deaths from lung cancer, ischaemic heart disease and chronic obstructive airways disease

- Passive smoking damages children before and after birth

- Thirty percent of pregnant women in Britain smoke

- Smoking is strongly related to poverty and deprivation

- Smoking costs the NHS an estimated £1.5 billion per year

- No other single avoidable factor accounts for such a high proportion of deaths, hospital admissions or general practitioner consultations

- Smoking is the single most important public health problem in Britain

Physical and pharmacological effects of nicotine

- Nicotine receptors are present in the brain and many other organs vary markedly in their binding, activation and desensitisation characteristics

- Cigarettes deliver rapid doses of nicotine to receptors in the brain

- Animal studies provide strong and consistent evidence that nicotine is addictive

- The addictive effect of nicotine is mediated at least in part by stimulation of dopamine release in the nucleus accumbens

- Pure nicotine has potential adverse effects on the human body but unlike cigarettes does not appear to cause cancer or significant cardiovascular disease

- Pure nicotine may be harmful to the fetus in pregnancy but is likely to be far less hazardous than the effects of smoking.

Psychological effects of nicotine and smoking

- Smoking is widely believed to have positive effects on mood
- Objective evidence suggests that the only improvements in mood resulting from smoking are those arising from the relief of withdrawal symptoms
- Smoking withdrawal symptoms are relieved by nicotine
- Nicotine intake in smokers is stable and consistent over time
- There is strong evidence of psychological dependence on cigarettes
- The major psychological motivation to smoke is the avoidance of negative mood states caused by withdrawal of nicotine

Is nicotine a drug of addiction?

- Nicotine obtained from cigarettes meets all the standard criteria used to define a drug of dependence or addiction
- Historically, and in contrast to addiction to opiates or alcohol, addiction to nicotine has not been recognised as a medical or social problem in Britain
- Nicotine is highly addictive, to a degree similar or in some respects exceeding addiction to 'hard' drugs such as heroin or cocaine
- Most smokers do not smoke out of choice, but because they are addicted to nicotine

The natural history of smoking: the smoker's career

- Addiction to nicotine is established in most smokers during teenage years, in many cases before reaching the age at which it is legal to buy cigarettes
- Teenagers who smoke one or more cigarettes per day demonstrate evidence of addiction similar to that seen in addicted adults, but addiction can be evident at lower levels of smoking
- Addiction to nicotine is usually established in young smokers within about a year of first experimenting with cigarettes
- A small proportion of smokers, approximately 5%, do not appear to be addicted to nicotine
- Once addicted, most smokers are unable to give up smoking even when they develop disease caused by smoking and made worse by continued smoking
- Only about 2% of smokers succeed in giving up in any year
- About 50% of young adult smokers will still be smoking when they are 60

Regulation of nicotine intake for smokers, and implications for health

- Smokers tend to regulate or titrate their nicotine intake to maintain body levels within a preferred range

- Smokers who switch to cigarettes which on machine smoking deliver less nicotine and tar tend to compensate for this by smoking the cigarette more deeply or more intensively

- Smokers of low yield cigarettes actually achieve little, if any, reduction in intake of nicotine and tar, and the health benefit accrued from switching to such cigarettes is, if anything, small

- The availability of low yield cigarettes may actually be counter productive in public health terms if they encourage health conscious smokers to switch to low yield brands instead of giving up completely

Management of nicotine addiction

- Effective interventions to reduce nicotine addiction are available at both population and individual levels

- The fact that smoking is so common in Britain means that even interventions that have small effects on smoking prevalence can, if widely applied, yield substantial returns in terms of the numbers of people who give up smoking

- Nicotine replacement therapy approximately doubles the effectiveness of most other currently available smoking cessation interventions

- Smoking cessation interventions, including nicotine replacement therapy, are extremely cost effective, costing society between £212 and £873 per year of life saved in 1996 prices

- The cost-effectiveness of smoking cessation interventions using nicotine replacement therapy compares very favourably with most other medical interventions

- Effective smoking cessation services should therefore be universally available to smokers through the NHS

- Smoking cessation services must be able to adapt to accommodate new effective therapies and interventions in the future

- Further research into the use and safety of nicotine replacement therapy relative to continued smoking during pregnancy is needed

Regulatory approaches to tobacco products in Britain

- Cigarettes are extremely damaging to consumers and yet have enjoyed unparalleled freedom from consumer protection regulation

- Much of the regulation applying to tobacco in Britain has been in the form of 'voluntary agreements' with the tobacco industry

- The use of additives in cigarettes has not been subject to appropriate assessments of public health impact

- The policy of progressively reducing tar yields from cigarettes, and of printing tar yields on cigarette packs, is based on flawed measurement methodology and may be ineffective in terms of achieving public health benefits

- Pharmaceutical nicotine delivery products (eg nicotine replacement therapy) are subject to regulation by the Medicines Control Agency and are required to meet the same safety standards as any other drug; however, cigarettes are exempt from these controls

- Cigarettes are tobacco-based nicotine delivery products and should be subject to the same safety standards as any other drugs

- A co-ordinated nicotine regulation framework needs to be established in Britain to resolve anomalies in the sale and promotion of nicotine delivery products, to maximise current and future public health

Main conclusions

- Most smokers do not continue to smoke cigarettes out of choice, but because they are addicted to nicotine

- Nicotine addiction is the underlying cause of the massive burden of premature death and disability caused by smoking in Britain

- Doctors, other health professionals and indeed society as a whole, need to acknowledge nicotine addiction as a major medical and social problem

- Treatment for nicotine addiction should be universally available for all smokers as a routine facility of the National Health Service

- Tobacco products must be made subject to safety regulations that are consistent with the controls that apply to all other drugs available in Britain, and so that they are commensurate with the extent of the damage to individuals and society that smoking causes

1 | Tobacco smoking in Britain: an overview

1.1 History of tobacco use in Britain

Earliest records

Tobacco is a native plant of the American continent. Historians believe tobacco began growing in the Americas around 6000 BC, and that American Indians started to use tobacco as early as the 1st century BC for medicinal and ceremonial purposes. The first pictorial record of tobacco being smoked was found on Guatemalan pottery dating from between the 7th and 11th centuries AD. By the time Europeans arrived on the American continent in the 15th century, smoking of tobacco among indigenous American people was widespread,[1] and native tribes were not only growing and consuming tobacco but also trading tobacco leaves.[2] Christopher Columbus was given tobacco, among other gifts, by American Indians in October 1492. Tobacco seeds and leaves were then brought back to Europe for the first time.

16th and 17th centuries: the age of the pipe

Tobacco was first introduced into English society in 1565 by Captain Sir John Hawkins, though Sir Walter Raleigh is more widely credited with making smoking fashionable in Britain some 20 years later.[1] In the 16th and 17th centuries, tobacco was commonly smoked in pipes, and although initially a prerogative of the rich, the habit gradually spread to all sections of society. Tobacco smoking remained almost exclusively a male habit, at least in public, until the 19th century.

The early growth in popularity of smoking tobacco was largely due to its supposed healing properties.[1] In 1571, a Spanish doctor, Nicholas Monardes, wrote a book on the history of medicinal plants of the New

World in which he claimed that tobacco could cure 36 health problems, including toothache, worms, lockjaw and cancer. However, by the early 1600s the reported benefits of smoking were beginning to be questioned, most notably by King James I (James VI of Scotland) who, in 1604, produced a damning report entitled 'Counterblaste to Tobacco', in which he said that smoking is a

> custome loathsome to the eye, hateful to the nose, harmful to the brain, [and] dangerous to the lungs.[1]

To discourage the smoking habit, King James increased the import tax on tobacco by 4,000%, and consumption fell dramatically as a result – but, after realising the effect of such a punitive level of duty on revenue from the tobacco trade, he subsequently reduced this tax and consumption rose again.

The first suggestion that tobacco smoking might be addictive was reported in 1610 when Sir Francis Bacon noted that trying to quit the habit was very difficult.[1] Some 70 years later, a lawyer, John Selden, observed that, just as some dislike sermons but might learn to enjoy them, the same applied to

> that which is the great pleasure of some men, tobacco; at first they could not abide it, and now they cannot be without it.[2]

Commercial cultivation of tobacco in America began in the early 17th century, and the first successful shipload of the new Virginian tobacco was sent to England in 1616.[1] In 1619, London clay pipe makers formed a chartered body. In the following year, a trade agreement between the Crown and the Virginia Company banned commercial growing of tobacco in England in return for a duty on imported Virginia tobacco of 1 shilling per lb weight.[1] By the late 1620s, an estimated 500,000 lb of tobacco were being brought into London every year, with smaller amounts going into the west coast ports,[2] and by the mid-1660s trade in tobacco between America and Europe had became a major business.[2] Tobacco use was now common among all sectors of society, and tobacco was sold in taverns, apothecaries' and tobacconists' shops. It was mostly in taverns that poorer people learnt to smoke, usually by sharing a communal pipe.

The 18th century: the age of snuff

In 1730, the first American tobacco factories were constructed, and in 1760 Pierre Lorillard established a factory in New York producing pipe tobacco, cigars and snuff.[1] During the 18th century, the trade in tobacco between Britain and America escalated, such that by 1770 approximately 96,000 hogsheads of tobacco (each representing a gross weight of around 600 kg) were imported into Britain each year, of which Britons consumed around 14,000 hogsheads. The rest was re-exported to Continental Europe, thus establishing England's role as a major tobacco trade centre. Following the restoration of the monarchy in 1660, the courtiers of Charles II returned to London from exile in Paris, and brought the French court's practice of snuff taking with them. For a period thereafter, snuff gradually replaced pipe smoking as the aristocratic form of tobacco use.[1]

While the demand for tobacco grew inexorably, some physicians began to warn of the potential dangers. In 1701, Nicholas Andryde Boisregard warned that young people taking too much tobacco have trembling, unsteady hands, staggering feet and suffer a withering of 'their noble parts'. In 1761, Dr John Hill performed possibly the first clinical study of tobacco's effects, and noted that snuff users were vulnerable to cancers of the nose. Thirty years later, he again reported cases in which snuff use had caused nasal cancer. In 1795, Samuel Thomas von Soemmering of Maine reported on cancers of the lip in pipe smokers.[1]

The 19th century: the age of the cigar

By the end of the 18th century, snuff-taking was in decline and a revival in smoking had begun, this time in the form of cigars rather than pipes. The fashion for cigars had spread from Spain, and with it came a smoking etiquette amongst the upper classes, whereby smoking tended to be confined to certain parts of the house, and gentlemen generally did not smoke in female company. Smoking by ladies was still frowned upon, though working-class women were known to use snuff or smoke clay pipes.[2] In 1826, the pure form of nicotine was discovered. By the 1860s, as smoking grew in popularity, it became more acceptable to smoke in public places. English railways introduced smoking carriages on trains, and public houses opened 'smoking saloons'.[2] Between the 1830s and 1870s, annual tobacco consumption almost doubled from 14 oz to 24 oz (ca 400–700 g) per head.[2] Although popular among the merchant and upper classes, smoking cigars remained an expensive habit, and poorer people still tended to smoke pipes or use snuff.[2]

It was the invention of the manufactured cigarette that transformed tobacco smoking into a truly mass habit. The origins of the cigarette lay in the Crimean war (1853–1856) when British soldiers copied the habit of hand-rolling tobacco from their Turkish allies. In 1854, London tobacconist Philip Morris began making hand-rolled cigarettes. The first cigarette factory opened in England in 1856, followed in 1871 by the establishment of Wills factory in Bristol and in 1888 by Player's in Nottingham. At the same time in the US, James 'Buck' Duke, the founder of American Tobacco, entered the cigarette manufacturing business.[1]

The 20th century: the age of the cigarette

James Duke was the first tobacco entrepreneur to use cigarette-making machines, and thus to initiate a revolution in the scale of cigarette production. In 1890, Duke formed The American Tobacco Company and set about taking over rival tobacco companies. He also set his sights on Britain, where a number of successful businesses such as Wills of Bristol had begun to use cigarette machines, producing around 85,000 cigarettes a day.[3] In 1901, in response to the threat of an American take over, 13 British tobacco companies joined together to form the Imperial Tobacco Company. A year later, The American Tobacco Company and Imperial came to an agreement to stay in their own countries and unite to form the British American Tobacco company to sell both companies' brands abroad.[3] Sales of manufactured cigarettes increased so rapidly from 1895 that by 1919 they accounted for more sales by weight than all other forms of tobacco combined.[4] Cigarette smoking among men increased further during World War I when cigarettes were included in soldiers' rations. As a result, many of the soldiers who returned home had become regular cigarette smokers.

It remained socially unacceptable for women to smoke until the Suffragette movement in the 1920s, when significant numbers of women started smoking.[5] The tobacco industry then began to market brands to women, using imagery associated with power and liberation; as a result, cigarette smoking escalated rapidly among both sexes during the 1930s and 1940s. By this stage, the British tobacco smoking epidemic was in full force.

1.2 Trends in smoking prevalence in Britain

The prevalence of tobacco smoking in Great Britain has changed dramatically during the 20th century. No direct measures of smoking prevalence are available for the first half of the century, but tobacco industry sales data demonstrate a greater than twofold increase in consumption of tobacco products during this period, from 4.1 g per adult per day in 1905 to a peak of 8.8 g in 1945 and 1946.[4] These peak levels of consumption have since been equalled (in 1959) but never exceeded.[4] The first available measure of smoking prevalence from tobacco industry sources is for 1948, when consumption was at 7.1 g per adult per day. At that time, an estimated 65% of men and 40% of women in Britain were regular smokers of manufactured cigarettes, and a further 16% of men smoked other tobacco products.[4] Industry figures suggest that male smoking declined progressively over the next two decades, whilst female smoking rose to a peak of 44% in 1966 and again in 1969 (Fig 1.1).

Since 1972, the prevalence of smoking in Britain has been measured regularly and independently from the tobacco industry in nationally representative population samples as part of the General Household Survey (GHS).[6] In the first year of the GHS, 52% of men and 41% of women in Britain were regular cigarette smokers, and a further 13% of men were smoking other tobacco products.[6] The prevalence of smoking then declined progressively in both sexes until 1994, when 28% of men and 26% of women were regular cigarette smokers. However, in 1996 (at the time of writing, the latest year for which GHS data have been published), the estimate of smoking prevalence had increased again in both men and women to 29% and 28%, respectively – the first increase in estimated smoking prevalence observed in Britain in either sex for 30 years (Fig 1.1).[6] Smoking prevalence estimates for late 1997 are available from the Omnibus Survey,[7] which in previous years has produced figures similar to those from the GHS. These data suggest that by 1997 smoking prevalence had fallen again in both men and women to 26% and 27%, respectively. Preliminary unpublished data for 1998, released by the Office for National Statistics shortly before going to press, give prevalence estimates of 28% for men and 26% for women.

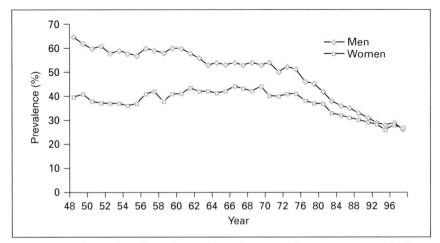

Fig 1.1. *Prevalence of smoking of manufactured cigarettes in men and women in Great Britain, 1948–1997 (Source: 1948–1971 Tobacco Advisory Council data;[4] 1972–1996 General Household Survey;[6] 1997 Omnibus Survey[7]).*

Overall, these estimates suggest that the prevalence of smoking in Great Britain may have been beginning to stabilise in recent years at a level of about one in four adults. However, inspection of age-specific rates in adults reveals that smoking prevalence has in fact been stable or increasing in recent years amongst most of the younger age groups, particularly in women (Fig 1.2). Similarly, inspection of trends in smoking amongst 15 year olds in England also shows a progressive increase during most of the 1990s,[8] again more marked in females (Fig 1.3). The rate at which new smokers are joining the prevalent smoking population has therefore been increasing for some years, making it likely that unless cessation rates also begin to increase, the overall prevalence of smoking in Great Britain will soon again begin to rise. Trends in smoking cessation rates can be inferred from trends in the proportion of ex-smokers amongst UK

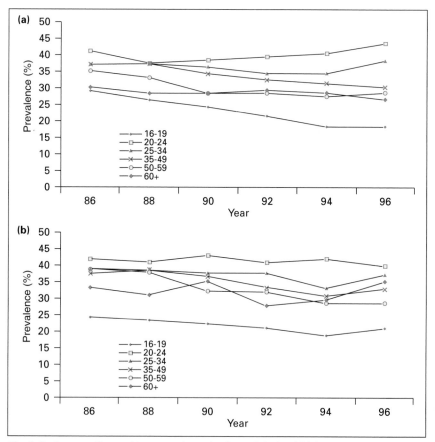

Fig 1.2. *Age-specific smoking prevalence in Great Britain, 1986–1996: (a) males; (b) females* (*Source*: General Household Survey[6]).

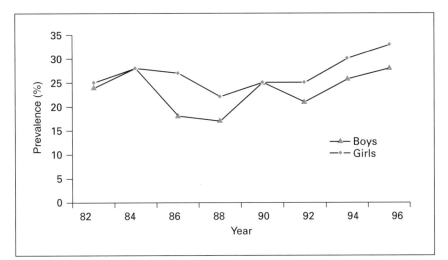

Fig 1.3. *Prevalence of regular smoking (at least one cigarette per week) amongst 15 year old boys and girls in England, 1982–1996* (*Source*: Office for National Statistics smoking among secondary schoolchildren surveys, 1982–1996[8]).

adults. These showed a progressive increase between 1970 and 1990, but have since remained remarkably stable (Fig 1.4). This suggests that cessation rates are no longer increasing, and that the prevalence of cigarette smoking in Britain is indeed in danger of increasing again over the next few years.

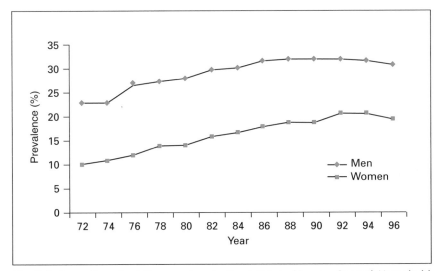

Fig 1.4. *Proportion of adult ex-smokers in Great Britain* (*Source*: General Household Survey[6]).

1.3 Risk factors and determinants of cigarette smoking

Gender

The effect of gender on the likelihood of being a smoker is changing. As described above, smoking in Britain has been more common in men for most of this century, but the difference in prevalence between men and women has been decreasing for many years and in 1997 the point estimate of cigarette smoking prevalence was actually slightly higher in women than in men.[7] This estimate did not allow for cigar or pipe smoking, so overall in Britain men are probably more likely than women to be smokers, but the difference between the sexes is now very small.

A trend towards female smoking has been evident for several years amongst schoolchildren, and the gap between the sexes has been increasing (Fig 1.5).[8] For some time therefore, females have accounted for the majority of young smokers entering the smoking population. Although the future relative prevalence of smoking in young male and female adults will depend on uptake and cessation rates in both sexes during the later teenage years, these data indicate that the proportion of females in the smoking population may be set to increase further.

Age

Age is a major determinant of smoking behaviour. Smoking is very uncommon in children up to and including the age of 11 years, but

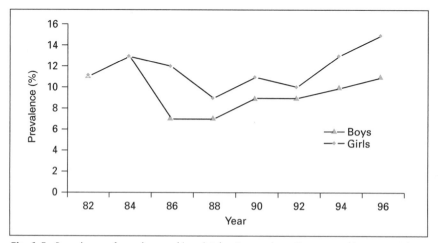

Fig 1.5. *Prevalence of regular smoking (at least one cigarette per week) amongst boys and girls aged 11–15 in England, 1982–1996 (Source:* Office for National Statistics smoking among secondary schoolchildren surveys, 1982–1996[8]).

increases substantially at 12–15 years old, to the extent that in 1996 28% of boys and 33% of girls were regular smokers by age 15 (Fig 1.6).[8] Amongst adults, smoking prevalence is greatest in the 20–24 age group (Fig 1.2), thereafter decreasing progressively with age.

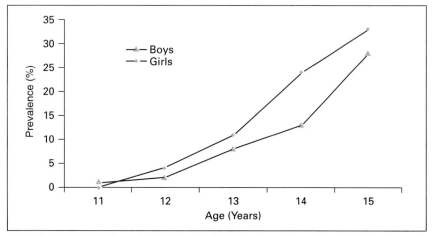

Fig 1.6. *Prevalence of regular smoking (at least one cigarette per week) with increasing age from 11–15 amongst boys and girls in England in 1996 (Source:* Office for National Statistics smoking among secondary schoolchildren survey, 1996[8]).

Socio-economic status

Smoking behaviour is strongly related to socio-economic status. In relation to occupation in 1996, smoking prevalence was lowest in the professional (12%) and highest in the semi-skilled manual occupational groups (39%) (Fig 1.7).[8] Data on the trend in smoking prevalence within non-manual and manual occupational groups suggest that the difference between these groups has, if anything, widened in recent years, more so in women (Fig 1.8). However, other measures of relative poverty or deprivation, including housing tenure, crowding, living in rented accommodation, being divorced or separated, unemployment, low educational achievement, and in women, single parent status, are also independently associated with an increased risk of smoking amongst adults.[9] Analysis of trends in smoking based on a composite index of some of these measures indicates that over the period 1973–1996 smoking prevalence fell by more than 50% in the most advantaged sector of British society, but has remained unchanged in the most deprived group.[9] Similar findings apply to smoking cessation rates, which also show a strong inverse relation with

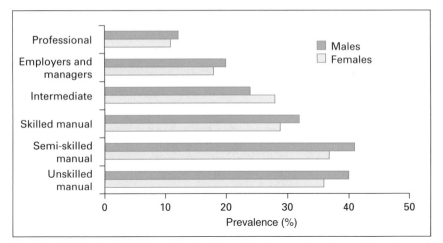

Fig 1.7. *Prevalence of regular smoking by occupational group in males and females aged 16 or over in 1996 (Source*: General Household Survey 1996[8]).

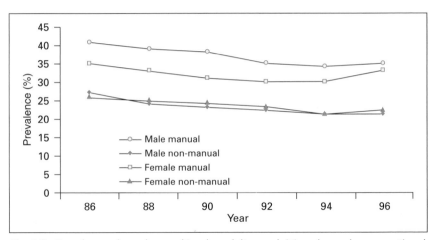

Fig 1.8. *Prevalence of regular smoking by adults aged 16 and over by occupational group and gender in England, 1982–1996 (Source*: General Household Survey 1986–96[8]).

deprivation. Cessation rates have doubled in the most advantaged groups, but have remained almost unchanged over the past two decades in the most disadvantaged sectors of society.[10]

Region of residence

Smoking prevalence varies in the regions. Data for NHS Regional Office areas of England reveal that the highest prevalence is in the North West Region (30%) and the lowest in the South and West Region (25%) (Fig 1.9).[8]

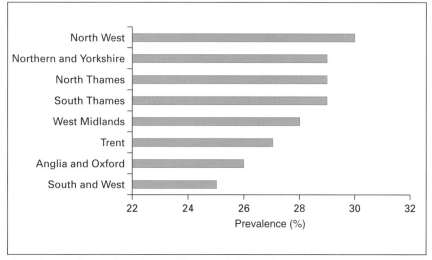

Fig 1.9. *Prevalence of regular smoking by adults aged 16 and over by NHS Regional Office area of England, 1996 (Source: General Household survey[8]).*

Risk factors for smoking in children

The factors associated with smoking in children broadly reflect those established for adults. Recent survey data[11] from children aged 11–15 years in England identify several factors associated with the likelihood of smoking in children, including:

- *low educational achievement*: children who are planning to take GCSE examinations, but with an expectation of passing in fewer than five subjects, are more than twice as likely to be smokers (26%) than those with higher expectations (10%);

- *living with parents who smoke*: children living with two parents who both smoke are nearly three times as likely to be smokers than those whose parents do not smoke, an effect particularly marked in girls (Fig 1.10);

- *having siblings who smoke*: children who have at least one sibling who smokes are four times more likely to smoke (26%) than those with no siblings who smoke (6%).

The following have been identified as additional potential risk factors:

- low socio-economic status
- having friends who smoke
- having teachers who smoke.[12]

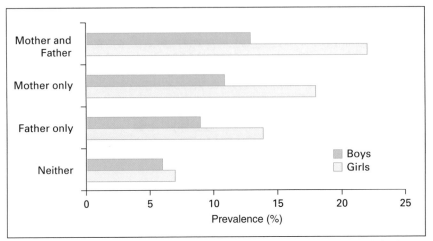

Fig 1.10. *Prevalence of regular smoking by children aged 11–15 who live with both parents, according to parental smoking habit, in England, 1997 (Source:* Teenage Smoking Attitudes Survey 1997[11]*).*

1.4 Smoking in pregnancy

Smoking during pregnancy is a problem of particular importance because of the harm that maternal smoking causes to the unborn child. The adverse effects of smoking during pregnancy have been reviewed in detail elsewhere,[12,13] but include spontaneous abortion, preterm birth, low birth weight and stillbirth. The children of mothers who smoke during pregnancy are at increased risk of neonatal mortality or sudden infant death syndrome, of asthma and/or wheezing illness in the first years of life, and they subsequently experience impaired physical growth and academic attainment compared to children of non-smoking mothers. These adverse effects are all imposed involuntarily and are, in principle, entirely avoidable.

Trends in smoking in pregnancy

In contrast to the overall trends in smoking in the general population, surveys of pregnant women conducted since 1992 by the Health Education Authority (HEA) in England suggest that smoking levels in this population have remained virtually unchanged. Published data from these surveys from 1992 to 1997 demonstrate little overall change in prevalence,[14] but subsequent data for 1998 and 1999 (HEA, data previously unpublished) indicate that the prevalence of smoking in pregnancy may be rising (Fig 1.11). In 1999, 30% of pregnant women were smokers.

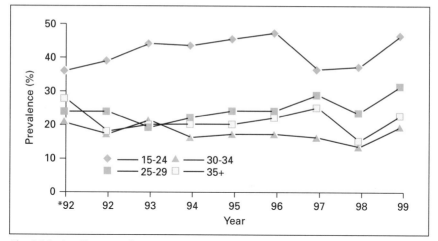

Fig 1.11. *Smoking prevalence among pregnant women in England by age, 1992–1999* (*Source*: Health Education Authority,[14] plus unpublished data for 1998–99). *The first measurement was made in January 1992, the second in March 1992; all other readings were made in March of the following years.

Determinants of smoking in pregnancy

Smoking during pregnancy is associated with many factors, particularly age, social class, education, marital status, presence of other smokers in the home, high parity, employment and ethnicity. An analysis of combined data from the HEA surveys described above confirms that smoking is twice as common amongst 15–24 year old pregnant women (42% smokers) than in those aged 35 and over (21% smokers), and that women in unskilled manual or unemployed groups were nearly six times more likely to smoke than those in professional and non-manual groups (45% and 8%, respectively) (Fig 1.12). Women who left full-time education at an early age were also much more likely to continue to smoke during pregnancy than other women (Table 1.1).

Relative to pregnant women as a whole, single/separated/divorced pregnant women and pregnant women who cohabit were also more likely to smoke (27%, 51% and 42%, respectively). In contrast, married women were less likely to smoke during pregnancy (17%) than pregnant women as a whole. Importantly, pregnant women with partners who smoke were four times more likely to be smokers themselves than were those with non-smoking partners (49% and 11%, respectively; 1999 survey). Family size was also associated with smoking during pregnancy: 23% of pregnant women with no children or one child were smokers, compared with 31% of pregnant women with two children, 37% with three children, and 45% with four or more children. Housing

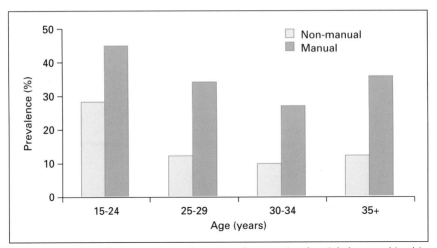

Fig 1.12. *Smoking during pregnancy by age and occupational social class combined in England, 1992–1999* (*Source*: Health Education Authority,[14] plus unpublished data for 1998–99).

Table 1.1. Relationship between age at which full-time education ended and smoking in pregnancy (data from Ref 14 and unpublished data for 1998–99).

Age at which full-time education ended (years)	% who smoked during pregnancy
15	48
16	34
17–18	15
19–20	11
21 or over	5

tenure was also related to smoking status, with 51% of pregnant women in rented council accommodation smoking compared with 18% of those who own their housing.

Smoking cessation in pregnancy

Although a significant number of women continue to smoke during pregnancy, many do make changes to their smoking behaviour around this time. The 1999 HEA survey found that 10% of women who were smoking immediately before pregnancy stopped and 4% cut down immediately beforehand. During pregnancy, however, more cut down than quit (33% and 20%, respectively). Even amongst those who managed to stop smoking either immediately before or during pregnancy, 19% had relapsed whilst still pregnant. These percentages have changed little over the last seven years.

The above findings show that smoking during pregnancy is a problem, particularly for women who are young, unemployed or from manual occupational groups, who left full-time education at a young age or are living in rented accommodation. The pattern of smoking in pregnancy therefore broadly reflects that in the general population, and demonstrates that interventions intended to reduce the prevalence of smoking in pregnant women, as in the general population, need to be directed especially towards socially disadvantaged groups. Recent trends in prevalence suggest that without new and effective interventions on smoking in pregnancy the target set in the recent White Paper, *Smoking kills*, is unlikely to be met:

> to reduce the percentage of women who smoke during pregnancy from 23% to 15% by the year 2010; with a fall to 18% by the year 2005.[15]

On a more positive note, however, many pregnant women do attempt to change their smoking behaviour, particularly in the very early stages of pregnancy. This highlights the potential for supporting pregnant women to stop smoking – and stay stopped – provided that adequate resources are made available.

1.5 Morbidity and mortality caused by smoking

Cigarette smoking causes substantial mortality and morbidity, the extent of which has been reviewed extensively elsewhere,[16-18] most recently for mortality in the UK in 1995 by Callum.[19] To provide more recent estimates of mortality and morbidity for the purposes of this report, these analyses have been repeated using updated 1997 UK mortality data and 1996 smoking prevalence data obtained from the sources previously described.[19] A similar approach has also been applied to estimate UK hospital admissions and general practitioner (GP) patients consulting in 1997–98 (April 1997–March 1998) from data provided to us by the Department of Health, the Office for National Statistics and the General Practitioner Research Database, and from equivalent sources in Wales, Scotland and Northern Ireland.

Deaths from smoking

In 1997, cigarette smoking accounted for an estimated 117,400 of the total of 628,000 deaths in the UK. Cigarette smoking is thus responsible for approximately one in every five deaths in Britain. This annual mortality translates into an average of 2,300 people killed by smoking every week, 320 every day and 13 every hour. The proportional impact of smoking is greater in younger age groups and in men, accounting for one in three male deaths in the 35–64 age group (Fig 1.13).

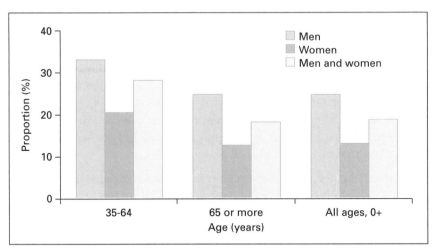

Fig 1.13. *Deaths due to smoking as a proportion of all deaths by age and sex, UK 1997.*

Years of life lost

Numbers of deaths do not convey a full sense of the loss to the community caused by smoking, since this depends not only on how many people die, but also how 'premature' their deaths are. Based on the distribution of deaths from smoking by age and mortality risks in never-smokers, we estimate that in 1997 cigarette smoking accounted for the loss of 205,000 years of life under age 65 and 551,000 years of life under age 75.

Deaths from smoking by disease

The number of deaths from individual causes, and the percentage of deaths attributable to smoking by cause for 1997 are summarised in Table 1.2. More than half of all smoking related deaths were due to respiratory disease (Fig 1.14). Smoking caused the majority of deaths from lung cancer, accounting for 89% of deaths from this disease among men and 74% among women. Similarly, 86% of deaths from chronic obstructive pulmonary disease (COPD) for men and 80% in women were attributable to smoking, as were approximately 17% of deaths from pneumonia. Cigarette smoking caused an estimated 17% of deaths from ischaemic heart disease, and 10% from stroke.

Table 1.2. Estimated number and percentage of deaths attributable to smoking by cause, UK 1997.

| | Number | | | Deaths from disease estimated to be caused by smoking | | |
| | | | | As % of all deaths from disease | | |
	Men	Women	Total	Men	Women	Total
Diseases caused in part by smoking						
Cancer						
Lung	19,600	9,600	29,200	89	75	84
Upper respiratory	1,500	400	1,900	74	50	66
Oesophagus	2,900	1,700	4,600	71	65	68
Bladder	1,600	300	1,900	47	19	37
Kidney	700	100	800	40	6	27
Stomach	1,600	300	1,900	35	11	26
Pancreas	600	900	1,500	20	26	23
Unspecified site	2,400	600	3,000	33	7	20
Myeloid leukaemia	200	100	300	19	11	15
Respiratory						
Chronic obstructive lung disease	14,000	9,700	23,700	86	81	84
Pneumonia	5,600	4,800	10,500	23	13	17
Circulatory						
Ischaemic heart disease	16,800	7,500	24,300	22	12	17
Cerebrovascular disease	3,000	3,800	6,900	12	9	10
Aortic aneurysm	3,800	2,000	5,800	61	52	57
Myocardial degeneration	200	300	500	22	12	15
Atherosclerosis	100	100	200	15	7	10
Digestive						
Ulcer of stomach or duodenum	900	1,000	2,000	45	45	45
Total caused by smoking	*75,600*	*43,200*	*118,800*			
Diseases prevented in part by smoking						
Parkinson's disease	900	400	1,300	55	28	43
Endometrial cancer	–	100	100	–	17	17
Total prevented by smoking	*900*	*500*	*1,400*			
Deaths from all causes due to smoking (caused less prevented)	**74,700**	**42,700**	**117,400**			

Totals may not add up due to rounding to nearest 100.

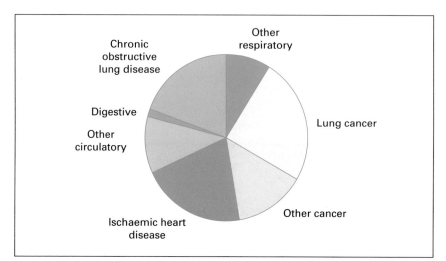

Fig 1.14. *Proportion of deaths attributable to smoking by disease, UK 1997.*

The estimated total of 117,400 deaths is a net figure which takes account of the small number of deaths from diseases prevented by smoking, principally from Parkinson's disease and endometrial cancer. Altogether, an estimated 118,800 deaths were caused by smoking and 1,400 deaths prevented. In 1997, therefore, smoking caused 85 times more deaths than it prevented.

Mortality and the individual smoker

Life-tables by smoking status, based on UK 1997 death rates and attributable percentages, can be constructed to portray the burden for the individual smoker. By this approach, a 35 year old man who smokes cigarettes can expect to die, on average, more than seven years earlier than a man who has never smoked (Fig 1.15). Cigarette smoking shortens life expectancy at age 35 years for women by six years compared with a woman who has never smoked cigarettes. Even for those who survive to age 65, cigarette smoking curtails their expected lifespan by more than six years among men and 5.5 years among women. The figures for ex-smokers lie between the two, closer to never-smokers than to current smokers. More than one in four men aged 35 who continue to smoke cigarettes can expect to die before age 65 compared to one in nine of never–smokers. The equivalent estimates for women are one in seven and one in 12, respectively. Overall, approximately one in every two smokers (51% of males and 45% of females) will die prematurely as a result of their smoking.

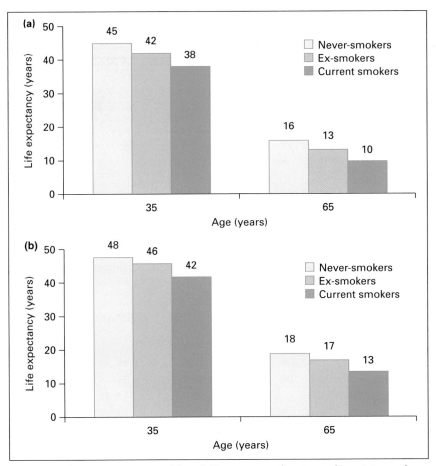

Fig 1.15. *Life expectancy at age 35 and 65 years according to smoking status and sex, UK 1997: (a) males; (b) females.*

Admissions to hospital

In 1997–8, an estimated 364,000 hospital admissions in England were attributable to the diseases caused by smoking listed in Table 1.3. This translates into 7,000 hospital admissions per week, or 1,000 per day. The admissions are spread fairly evenly across the main groups of diseases caused by smoking (Fig 1.16):

- 109,000 from cancer
- 112,000 from respiratory disease other than lung cancer
- 134,000 from circulatory disease.

The major fatal diseases caused by smoking (listed in Table 1.2) were responsible for 6% of hospital admissions among those aged 35 years or over (9% for men and 4% for women). This understates

Table 1.3. Estimated number and percentage of UK hospital admissions attributable to smoking by cause, 1997–98.

	Hospital admissions* for diseases (as listed in Table 1.2)					
	Number			As % of all admissions for disease		
	Men	Women	Total	Men	Women	Total
Diseases caused in part by smoking						
Cancer						
Lung	33,300	17,100	50,400	90	77	85
Upper respiratory	8,100	2,400	10,400	75	53	69
Oesophagus	9,300	4,500	13,800	71	66	69
Bladder	18,700	2,900	21,600	47	19	40
Kidney	1,700	200	1,900	41	7	28
Stomach	4,800	700	5,500	35	11	28
Pancreas	1,100	1,300	2,400	21	28	24
Unspecified site	1,900	400	2,200	25	4	14
Myeloid leukaemia	1,300	600	1,900	19	11	15
Respiratory						
Chronic obstructive lung disease	53,000	42,800	95,800	86	82	84
Pneumonia	9,100	6,900	16,000	26	20	23
Circulatory						
Ischaemic heart disease	72,400	28,400	100,800	34	23	30
Cerebrovascular disease	11,400	11,500	23,000	19	18	18
Aortic aneurysm	6,500	1,900	8,400	62	55	60
Atherosclerosis	1,200	500	1,700	19	13	17
Digestive						
Ulcer of stomach or duodenum	7,700	6,400	14,100	49	51	50
Total caused by smoking	*241,400*	*128,500*	*369,900*			
Diseases prevented in part by smoking						
Parkinson's disease	3,200	1,300	4,500	56	30	45
Endometrial cancer	–	1,200	1,200	–	17	17
Total prevented by smoking	*3,200*	*2,500*	*5,700*			
Hospital admissions from all causes due to smoking (caused less prevented)	**238,200**	**126,000**	**364,200**			

*admissions with disease as primary cause.
Totals may not add up due to rounding to nearest 100.

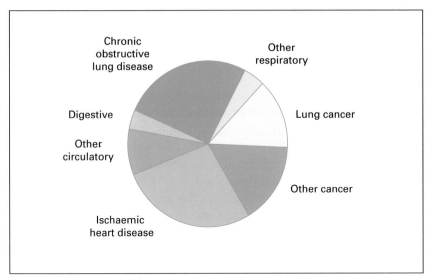

Fig 1.16. *Hospital admissions attributable to smoking by disease, UK 1997–98.*

the overall impact of smoking on hospital admissions to the extent that non-fatal diseases caused by smoking are excluded from these estimates.

General practitioner consultations

In primary care in 1997/8, cigarette smoking caused an estimated 480,000 patients to consult their GP for ischaemic heart disease, 20,000 for stroke and nearly 600,000 for COPD.

Conclusions

The burden of premature mortality and morbidity caused by smoking in Britain is massive. No other single avoidable cause of disease accounts for such a high proportion of deaths, hospital admissions or GP consultations. Cigarette smoking is the single most important public health problem in Britain.

1.6 The costs of smoking in Britain

The harmful effects of smoking tobacco can also be considered from an economic perspective and expressed in money terms or as costs. Such costs can include the effects on the individual as well as those borne by the wider community.

The health burden of smoking

One of the main costs of tobacco smoking is the health burden it creates. For most purposes, for example in the evaluation of new medical therapies, the loss of quality and quantity of life is not directly valued, rather measures such as quality adjusted life years are adopted. However, there are monetary equivalents with, for example, the Department of Transport putting the value of the human costs of a loss of life in 1997 at £680,590.[20] If this value per life is applied to the total number of deaths attributable to smoking in Table 1.2, the cost of smoking related mortality in the UK in 1997 prices is just under £80 billion. This estimate does not take account of the value to the individual of smoking related illness, or of related loss of quality of life. Clearly, while smokers know of the risks to health and may set aside some of these individual costs against the 'benefits' they receive from smoking, the majority of smokers wish to quit. Many smokers will spend considerable sums of money on a variety of smoking pro-grammes before successfully stopping. The human costs of smoking expressed in money terms are thus considerable.

The costs of smoking to society

The costs of smoking to the rest of society can be divided into:

- the costs of the harmful effects of passive smoke exposure in non-smokers
- the costs imposed by smokers on the wider community through, for example, the use of scarce health service resources or by lower productivity in the workplace.

Passive smoking. The health risks arising from passive smoking have a direct impact on family members, especially on children and the unborn, and in the workplace. As with the direct health effects on smokers, a high value could be placed on any premature loss of life or passive smoking related illness. There can be considerable financial consequences for the health and other welfare agencies from a pre-mature birth. Models of the cost-effectiveness of smoking cessation interventions with pregnant women indicate that these are potentially cost saving because of these high costs. Children living with smokers also have poorer health and higher health care expenditures than those in non-smoking households. Stoddard and Gray[21] estimated that passive smoking was responsible for 19% of all expenditures for child-hood respiratory conditions in the US. In Hong Kong, it has been estimated that the cost per child of GP consultations was 14% higher for children living with one smoker at home and 25% higher where there were two or more smokers.[22] Godfrey *et al*[23] estimated that the cost to the NHS of the effects of passive smoking on children was £410 million in England and Wales.

Fires caused by smoking are another effect which may affect both smokers and non-smokers. Buck and Godfrey[24] estimated the costs of fires at £150 million for England and Wales in 1991. This figure excludes any value on the loss of life from such fires.

Health service resources. Clearly, there are effects on the health services from smoking related illness. Two general approaches can be used to estimate these costs to the NHS:

- by attributing the costs of different diseases to smoking
- by estimating the different health care costs of smokers compared to non-smokers.

Parrott *et al*[25] used both methods to estimate the annual smoking related costs in England in 1996/1997 prices. In the first approach, estimates of smoking related, attributable fractions were applied to NHS costs by disease code. The total estimated cost was £1.5 billion for England. The breakdown by disease and type of NHS expenditure is shown in Table 1.4. The alternative method combined data on health care utilisation by current smokers and never-smokers, taken from the GHS with unit costs for different types of health care use from a variety of sources (see Ref 25 for details), and reached a similar total of £1.4 billion. The breakdown for different types of health service use estimated by this method are summarised in Table 1.5. The

Table 1.4. Estimated smoking related costs to the NHS by disease group, England 1996–1997.[25]

Main disease group	Hospital cost (£ million)	Primary care cost (£ million)	Pharmaceutical cost in primary care (£ million)	Total
Cancer	203	19	0*	222
Respiratory	273	72	0*	345
Circulatory	639	61	139	839
Digestive	100	4	0*	104
Total	1,215	156	139	1,509

Totals may not add up due to rounding.
*Separate primary care pharmaceutical cost estimates not available.

Table 1.5. Health care costs estimated from observed differences between smokers and non-smokers, England 1996–1997.[25]

Type of health service use	Cost (£ million)
General practitioner visits	250
Prescription costs	150
Inpatient stays	320
Day cases	190
Outpatient visits	490
Total	1,400

two methods yielded similar totals, but the second shows slightly higher primary care costs.

These types of figures for smoking related health care expenditure are based on annual estimates. It is more difficult to estimate how such figures may change over time if smoking prevalence declines. However, there is clearly a large potential for NHS resources currently being used to treat smoking related problems to be diverted to treating other problems. It is sometimes argued that, although smokers consume more health care throughout their lifetime, their life expectancy is shorter. Authors have attempted to calculate whether smokers or non-smokers have the higher total lifetime health care expenditures. Results from studies have varied (see, for example, Refs 26 and 27), and we have no estimate for the UK. The implications from such studies are not clear. There may be some fairness issues involved, for example, do smokers pay their way? Similar arguments surround the balance of tax paid by smokers and benefits received.[28]

There are, however, more conflicting views as to whether the health care consequences of 'unrelated' conditions should be included in any analysis of the worth of interventions such as smoking cessation programmes. Extending these arguments to all types of health care conditions would suggest that there are 'costs' both to any life pro-longing intervention and to most preventive measures. This would especially disadvantage interventions which save the lives of younger people, and it seems unlikely that such arguments would be upheld by society. Such an analysis obviously fails to acknowledge the benefits to society of people living longer. Also, if differential timing is taken into account by discounting or putting a lower weight on future events, the result would seem to indicate that smokers do have lifetime excess health care costs.[27] This suggests that such empirical estimates are a diversion from the main point that smoking cessation interventions can yield high health gains at low cost.

Another area where smokers can have a major impact is the workplace. Smokers, in general, have more sickness absences, and in workplaces where smoking breaks are allowed there can also be a considerable loss in productivity. Parrott *et al*[29] estimated that total productivity losses in Scotland due to smoking in the workplace are almost £400 million per year. Many smokers have to give up work because of smoking related illness. Buck and Godfrey[24] estimated that in England and Wales in the year to March 1991, 34 million working days were lost from smoking related illness, at an estimated cost of £328 million.

Money figures are one way of trying to summarise the wide-ranging effects of smoking on society. It is difficult to give a total UK annual burden of smoking. Figures quoted above are taken from different studies at different time periods and covering different parts of the UK. There are also some missing figures, for example, of the cost to the unborn. Also, opinions vary as to the items which should be included or excluded in such a total figure. However, irrespective of the effect of these considerations on the total figure, it is clear that cigarette smoking has major health and economic impacts in Britain.

References

1   Borio G. *Tobacco timeline, 1993–1998.* www.tobacco.org/History/Tobacco_History.html.

2   Kiernan VG. *Tobacco: a history.* London: Hutchinson Radius, 1991.

3   Taylor P. *The smoke ring: tobacco, money and multinational politics.* London: Sphere Books, 1985.

4   Nicolaides-Bouman A, Wald N, Forey B, Lee P. *International smoking statistics.* Oxford: Oxford University Press, 1993.

5   Vierola H. *Tobacco and women's health.* Helsinki: Art House Oy, 1998.

6   Office for National Statistics. *Living in Britain.* London: The Stationery Office, 1997.

7   Freeth S. *Smoking-related behaviour and attitudes, 1997.* A report on research using the Omnibus Survey produced on behalf of the Department of Health. London: Office for National Statistics, 1998.

8   *Statistics on smoking: England, 1976 to 1996.* Department of Health Bulletin 1998/25. London: Department of Health, 1998.

9   Jarvis MJ, Wardle J. Social patterning of individual health behaviours: the case of cigarette smoking. In: Marmot M, Wilkinson R (eds). *Social determinants of health.* Oxford: Oxford University Press, 1999: 240–55.

10 Jarvis MJ. Patterns and predictors of smoking cessation in the general population. In: Bolliger CT, Fagerström KO (eds). *The tobacco epidemic.* Basel: Karger, 1997: 151–64.

11 Barton J. *Young teenagers and smoking in 1997.* London: Health Education Authority, 1998.

12 *Smoking and the young.* A report of a working party of the Royal College of Physicians. London: RCP, 1992.

13   Charlton A. Children and smoking: the family circle. *Br Med Bull* 1996; **52**: 90–107.

14   Owen L, McNeill A, Callum C. Trends in smoking during pregnancy in England, 1992–7: quota sampling surveys. *Br Med J* 1998; **317**: 728.

15   *Smoking kills. A White Paper on tobacco.* London: The Stationery Office, 1998.

16   Doll R, Peto R, Wheatley K, Gray R, Sutherland I. Mortality in relation to smoking: 40 years' observations on male British doctors. *Br Med J* 1994; **309**: 901–11.

17   Peto R. Smoking and death: the past 40 years and the next 40. *Br Med J* 1994; **309**: 937–9.

18   Peto R, Lopez AD, Boreham J, Thun M, *et al.* Mortality from smoking worldwide. *Br Med Bull* 1996; **52**: 12–21.

19   Callum C. *The UK smoking epidemic: deaths in 1995.* London: Health Education Authority, 1998.

20   Department of Transport. *Valuation of the benefits of prevention of road accidents and casualties.* Highways Economic Note No 1, 1997. London: Department of the Environment, Transport and the Regions, 1998.

21   Stoddard JJ, Gray B. Maternal smoking and medical expenditures for childhood respiratory illness. *Am J Pub Health* 1997; **87**: 205–9.

22   Peters J, McCabe CJ, Hedley AJ, Lam TH, Wong CM. Economic burden of environmental tobacco smoke on Hong Kong families: scale and impact. *J Epidemiol Community Health* 1998; **52**: 53–8.

23   Godfrey C, Edwards H, Raw M, Sutton M. *The smoking epidemic: a prescription for change.* Health Education Authority: London, 1993.

24   Buck D, Godfrey C. *Helping smokers give up: guidance for purchasers on cost-effectiveness.* Health Education Authority: London, 1994.

25   Parrott S, Godfrey C, Raw M, West R, McNeill A. Guidance for commissioners on the cost–effectiveness of smoking cessation interventions. *Thorax* 1998; **53**(Suppl 5, Part 2): S1–38.

26   Hodgson TA. Cigarette smoking and lifetime expenditures. *Millbank Q* 1992; **70**: 81–125.

27   Barendregt JJ, Bonneux L, Van De Maas PJ. The health care costs of smoking. *N Engl J Med* 1997; **337**: 1052–7.

28   Cohen D, Barton G. The cost to society of smoking cessation. *Thorax* 1998; **53**(Suppl 2): S38–42.

29   Parrott S, Godfrey C, Raw M. *Cost and benefit analysis of smoking cessation in the workplace.* Final report to the Health Education Board for Scotland. University of York: Centre for Health Economics, 1996.

2 | Physical and pharmacological effects of nicotine

2.1 Nicotine receptors and subtypes

Introductory perspective: general features of nicotinic acetylcholine receptors

Although nicotine was first used by Langley at the turn of the century in his classic experiments that gave rise to the concept of a 'receptive substance', and hence 'receptor', it was many years before receptors for nicotine in the brain were identified. Langley, and later Dale (who distinguished the nicotinic and muscarinic actions of acetylcholine), examined nicotine on skeletal muscle or autonomic ganglia. The utility of muscle preparations for electrophysiological analyses, together with the model system of the electric organ from the marine ray *Torpedo* that provides a rich source of receptor protein, have resulted in the muscle nicotinic acetylcholine receptors (nAChR) being the best characterised ligand-gated ion channel to date.[1]

Briefly, the nAChR is a pentamer composed of five homologous membrane-spanning subunits around a central pore or ion channel. Two α subunits and one each of β, γ and δ subunits (with a change during development from γ to ε) are arranged in the order $\alpha\gamma\alpha\delta\beta$. The two α subunits are the primary agonist binding subunits, and the co-operative binding of two molecules of acetylcholine is required to open the channel through an allosteric mechanism. The ion channel of the muscle nAChR is primarily permeable to Na^+ and K^+. In normal physiological conditions, opening of the channel results in an inward flux of Na^+ producing local depolarisation that can lead to muscle contraction. The nAChR channel remains open for only a brief period before undergoing a series of conformational changes that produce a *desensitised* state (Fig 2.1). In this configuration, the channel is closed to ions and is refractory to activation by agonist, although

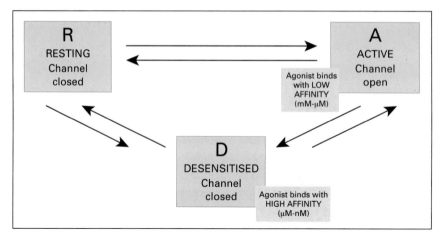

Fig 2.1. *Conformational states of the nicotinic acetylcholine receptor.*

agonist can still bind to the receptor with enhanced affinity. Low concentrations of agonist can push the receptor into the desensitised state without going through the open state. These properties have implications for the functional effects of nicotine during tobacco use.

Heterogeneity of neuronal nicotinic acetylcholine receptors

Neuronal nAChRs share the same overall structural and functional features of the muscle nAChR prototype, but generally comprise two α and three β subunits, and also involve different subunits to those expressed in muscle. The number of neuronal nAChR subunit genes identified to date (9 in mammals, including humans) could generate a vast number of pentameric combinations. Certain combinations are not viable, but there is nevertheless considerable heterogeneity (for review, see Ref 2). The different pharmacological and biophysical properties of these subtypes will have a number of implications for the actions of nicotine (Table 2.1), although the subunit composition and physiological function of any nAChR subtype in the brain is only poorly delineated (see below). Hypothetically, a smoker's average plasma level of nicotine sustained throughout the day may be sufficient to desensitise (partially or fully) a population of nAChR with high affinity for nicotine (also making it unresponsive to the natural ligand, acetylcholine), while the raised nicotine levels from a bolus of cigarette smoke will activate/desensitise other, less sensitive nAChR subtypes to varying degrees. The perceived effect of smoking (stimulation/relaxation) will be the balance between these processes.

Table 2.1. Implications of nicotinic acetylcholine receptor (nAChR) heterogeneity.

nAChR subtypes may vary with respect to:
- anatomical localisation (pathways)
- cellular localisation (presynaptic/nerve terminals or cell soma/dendrites)
- proximity to sites of ACh release (postsynaptic or extrasynaptic)
- sensitivity to nicotine (activation vs desensitisation)
- rate of desensitisation
- Ca^{2+} permeability (cell regulation, long-term potentiation)
- regulation (allosteric/phosphorylation)

Nomenclature of nicotinic acetylcholine receptor subunits

The nine neuronal nAChR subunit genes cloned to date from mammals and shown to be expressed in neurones are designated α ($\alpha 2$–$\alpha 7$) or β ($\beta 2$–$\beta 4$), with muscle α and β subunits being the first in the series. α subunits are characterised by the presence of a pair of adjacent cysteine residues close to the acetylcholine binding site in the N-terminal domain, and shown to be important for agonist binding. Hence, α subunits have also been referred to as 'agonist binding subunits', whereas β subunits have been called 'structural subunits' on the presumption that they contribute to the formation of the nAChR channel but play no role in agonist recognition. These distinctions are incorrect. The $\alpha 5$ subunit appears to be incapable of binding agonist because it lacks another key residue – a tyrosine found in all other α subunits upstream of the pair of cysteine residues mentioned above.[3] Moreover, β subunits do influence nAChR pharmacology: the agonist binding site is thought to reside at the interface between an α and the adjacent subunit. Thus, exchanging $\beta 4$ for $\beta 2$ in an $\alpha\beta$ hetero-oligomer expressed experimentally in a system such as *Xenopus* oocytes resulted in an increase in sensitivity to nicotine.[4] The naming of nAChR subunits and subtypes has recently been reviewed by a subcommittee of the International Union of Pharmacology Nomenclature Committee.[5]

Prevalence and distribution of nicotinic acetylcholine receptor subunits

Evidence from *in situ* hybridisation and immunocytochemistry indicates that the various nicotinic subunits have different, but overlapping distributions in the brain (the anatomical distribution in the human brain is reviewed by Gotti *et al*[6]). They differ in abundance, with $\beta 2$ the most widely expressed subunit. The loss of nicotine self-administration behaviour in knock-out mice lacking this subunit suggests that it contributes to nAChR relevant to nicotine dependence.[7] α7 and α4 are also widely expressed, but with complementary distributions: for example, there are high levels of α7 in hippocampus, but low expression of α4. The other subunits have more restricted distributions: for example, α6 and β3 gene transcripts are limited to areas containing dopamine and noradrenaline cell bodies,[8] and the α6 subunit has been localised to dopamine neurones in the ventral tegmentum (VTA).[9] Lack of coincident expression prevents certain subunit combinations from occurring, but the distribution patterns are compatible with enormous heterogeneity of nAChR subtypes. For example, the VTA expresses α3, α4, α5, α7 and β2 in addition to α6 and β3. Heterologous expression systems have demonstrated restrictions in the assembly of subunits to create functional nAChR (summarised briefly in Table 2.2).

Neuronal nicotinic subunits have also been reported to be expressed in certain non-neuronal cells, including lymphocytes, skin, epithelial cells and small-cell lung carcinoma.[6] The novel α9 subunit is expressed in sensory end organs, notably the outer hair cells of the auditory system.[10]

Subunit composition of native nicotinic acetylcholine receptors

The question then arises whether these nAChR formed in experimental systems represent native nAChR occurring in the brain. This is a much less tractable problem, with few definitive answers so far. α7-like nAChRs have pharmacological and biophysical properties and high Ca^{2+} permeability, comparable to homomeric α7 nAChR in expression systems, but there is debate about the possible inclusion of additional types of subunit.[11] More recently, data in chicken[12] and in rat cardiac

Table 2.2. Nicotinic acetylcholine receptor (nAChR) subtypes: evidence from heterologous expression systems.

- α7 forms robust, homomeric nAChR, whereas α2-α6, β2-β4 do not

- Pairwise combinations of α2, α3 or α4 with either β2 or β4 give functional nAChR

- α5 can be incorporated into nAChR with α3β4 or α3β2, and produces changes in channel properties and agonist sensitivity

- β3 may also contribute a third type of subunit to α/β combinations

- Until recently, rat/chick α6 only formed nAChR with human β2

ganglia[13] support variants of α7-like nAChR. Although α7 nAChRs are not particularly sensitive to nicotine (EC_{50} human α7 ~ 40 μM), it has been demonstrated *in vitro* that this subtype can enhance glutamate release from hippocampal nerve endings in response to 500 nM nicotine. This is interpreted as an ability to respond to the levels of nicotine in smokers.[14]

Immunoprecipitation studies provide good evidence for the occurrence of α4β2 hetero-oligomers.[15] This nAChR has the highest sensitivity to nicotine (EC_{50} ~ 0.1–1 μM), but its functional status in smoking is equivocal because it can be desensitised by lower concentrations.[16] Other nAChR subtypes/subunits are less abundant in the brain:

- α3α5β4/β2 nAChR mediate synaptic transmission in sympathetic ganglia
- α3/α4+β4/β2+α5 nAChR occur in the brain (for review, see Ref 2)
- nAChR containing α3 and β2 subunits can modulate dopamine release in striatum.[17]

The expression of α6 and β3 subunits predominantly in catecholaminergic nuclei, including the VTA, makes them candidate subunits of nAChR operative in the Reward pathway. Transgenic animals deficient in (or overexpressing) particular subunits will help to clarify their roles. To date, β2, α7, β3 and α4 knock-out mice have been reported, but so far only the β2 knock-out has been characterised with respect to rewarding and other behaviours.[7,18]

Pharmacological distinctions

There is a relative lack of specific ligands for studying nAChR. Nature has provided the most selective and potent tools, although the burgeoning interest in nAChR as a target by the pharmaceutical industry is generating promising new compounds for research purposes.

Radioligands. Radiolabelled ligands are useful for:

- quantifying and mapping the distribution of nAChR
- determining pharmacological profiles
- assaying nAChR during biochemical procedures such as purification.

Radioligands have defined two major populations of nAChR. The first is *α7-type nAChR.* The snake toxin αbungarotoxin (αBgt) is a specific and almost irreversible α7 antagonist. [^{125}I] or [^{3}H]αBgt labels this subtype with an affinity of about 1 nM. Following the use of [^{125}I]αBgt to label muscle nAChR, [^{125}I]αBgt binding sites were characterised in mammalian brain in the 1970s, but their receptor status was controversial at that time. Cloning and expression of the α7 subunit in 1990 confirmed that [^{125}I]αBgt labels an nAChR.[19] A tritiated version of the *Delphinium* toxin methyllycaconitine (MLA) has recently been developed,[20] which also selectively labels α7-type nAChR with an affinity of about 1 nM. However, its binding is reversible and it can discriminate between muscle and α7 nAChR.

The second defined population is *α4β2 nAChR.* Tritiated nicotine was first reported to bind to brain tissue by Romano and Goldstein in 1980.[21] Since then, [^{3}H]nicotine binding has been extensively characterised, and the binding sites appear identical to those labelled with other agonists, namely [^{3}H]acetylcholine, [^{3}H]cytisine, [^{3}H]methylcarbamylcholine and [^{3}H]ABT418. Immunoprecipitation experiments indicate that these sites correspond to α4β2 nAChR.[15] Consistent with this, [^{3}H]nicotine binding is absent in β2 knock-out mice.[18]

These radiolabelled agonists bind to brain tissue with affinities of about 1–10 nM. Affinities of this order are required for the successful use of a radioligand so, although by definition, all nAChR bind nicotine, only the α4β2 type appears to be sensitive enough for this binding to be measurable in an assay. It should be noted that the nanomolar binding affinities of these agonists are lower than their EC$_{50}$ values for activating α4β2 nAChR (0.1–10 μM), and reflect binding to the *high affinity, desensitised state* of the nAChR (Fig 2.1).[16] While competition binding assays with such agonists can provide information on the relative potencies of competing ligands, the results can be misleading with respect to receptor activation. Binding assays do not distinguish agonists from antagonists, and full inhibitors of binding can be partial agonists with respect to nAChR function. [^{3}H]nicotine has also been used for positron emission tomography studies in humans to examine the distribution and changes in numbers of nAChR.[22]

The portfolio of tritiated agonists was recently extended by the advent of [^{3}H]epibatidine, secreted from the skin of a South American frog. It is the most potent nicotinic agonist to date and binds to brain tissue with an affinity of about 10–100 pM.[23] As well as labelling the same nAChR as the other tritiated agonists discussed above (α4β2), [^{3}H]epibatidine also identifies one or more additional nAChR.[24] The persistence of [^{3}H]epibatidine labelling of the medial habenula and interpeduncular nucleus in β2 knock-out mice is interpreted in favour of an nAChR containing α3 and β4 subunits. [^{3}H]epibatidine labels α3-containing nAChR in sympathetic cell lines lacking the α4 subunit. Thus, [^{3}H]epibatidine is useful for labelling additional subtypes, but this can be problematic in tissue with a heterogeneous population of nAChR such as brain, where multiple subtypes are present.

The minor snake venom component, known as neuronal bungaro-toxin (nBgt), has been iodinated and used to label brain nAChR. In the presence of αBgt to block its binding to α7-type nAChR, [^{125}I]nBgt labels a smaller population of sites tentatively equated with α3-containing nAChR.[25] nBgt is not commercially available, and concerns about its purity and stability, together with loss of activity on iodination, have limited its utility.

Antibodies have also been valuable tools for histological localisation (eg Ref 9), isolation,[3] purification[11] and quantitation[26] of specific neuronal nAChR subtypes and subunits.

Functional studies: agonists. Although all nAChR, by definition, respond to nicotine, they differ with respect to the nicotine concentrations required for activation or desensitisation. Selected EC_{50} values for nicotine and acetylcholine of heterologously expressed human nAChR subtypes are compared in Table 2.3.

Table 2.3. EC_{50} values (μM) for nicotine and acetylcholine at nicotinic acetylcholine receptor subtypes.[6]

	α4β2	α3β2	α3β4	α7
Nicotine	5	132	80	40
Acetylcholine	68	440	203	79

No agonists are truly specific for a particular subtype of nAChR, but may differ in potency by one or two orders of magnitude. The tobacco alkaloid anabasine and its synthetic derivative GTS-21 (also known as DMXB) have diminished potency and efficacy at α4β2 nAChR in *Xenopus* oocytes, relative to α7 nAChR at which they are highly effica-cious. Thus GTS-21 has been referred to as *functionally selective* for α7.[27] This ignores its possible interactions with minor subtypes of nAChR. Choline is reputedly an α7-selective agonist.[28] Although it is not very potent (EC_{50} 1.6 mM), micromolar concentrations that might occur in the brain readily desensitise the α7 nAChR. Thus, actions of nicotine must be viewed against a backdrop not only of endogenous acetyl-choline but also of its principal metabolite. With regard to efficacy, it seems that the α4β2 nAChR is particularly prone to submaximal activation/partial agonism.[29]

Competitive antagonists. Competitive antagonists (Table 2.4) compete with agonists for binding to the same or overlapping sites (see Ref 6

Table 2.4. Selectivity and potency of some nicotinic receptor antagonists.

Antagonist	Source	Subtype selectivity	Potency IC_{50}
αBgt	*Bungarus multicinctus*	α7	1 nM
MLA	*Delphinium* sp	α7 (>α3>α4)	1 nM
αconotoxin IMI	*Conus imperialis*	α7 (>α9>α1)	220 nM
αconotoxin MII	*Conus magnus*	α3β2	~1 nM
αconotoxin AuIB	*Conus aulicus*	α3β4	~1 µM
DHβE	*Erythrina*	α4 (>α3>α7)	0.1 µM
mecamylamine	synthetic	α3 (>α4>α7)	0.1 µM

αBgt = αbungarotoxin; DHβE = dihydro-β-erythroidine; MLA = methyllycaconitine.

for a more comprehensive account). They can exhibit a high degree of nAChR subtype selectivity. Competitive antagonists acting at α4β2 nAChR displace [^{3}H]nicotine binding, while those acting at α7 nAChR displace [^{125}I]αBgt binding.

- *αBgt*, as already noted, binds specifically to α7-type nAChR and is diagnostic for the presence of this subunit (although it will also label nAChR composed of the α8 subunit, found only in chickens, and the α9 and muscle nAChR). Low nanomolar concentrations of αBgt block α7 responses. The blockade is reversible only very slowly, but its large size and slow kinetics can be a disadvantage. αBgt does not block other mammalian neuronal nAChR subtypes.
- *MLA* is α7-selective rather than specific. 1–10 nM concentrations will reversibly block α7-type nAChR, while 100 nM MLA begins to block α3-type receptors.[30] Concentrations above 1 µM are required to antagonise α4β2 responses.
- *Dihydro-β-erythroidine* (DHβE) preferentially blocks α4β2 nAChR at submicromolar concentrations that do not impair the function of α7 nAChR.
- *αConotoxins* (Table 2.4), peptides with nicotinic antagonist activity, are produced by marine cone snails. They have high potency and selectivity, and are proving very useful for *in vitro* studies.[31]

Non-competitive antagonists. Non-competitive antagonists do not bind at the agonist binding site. In many cases, they act by occluding the ion channel of the nAChR. If they enter the channel, their action will be voltage-dependent. Compounds which block the channel may lack specificity and also block non-nicotinic receptor channels.

- *Mecamylamine* is the classical nicotinic antagonist, much used for *in vivo* studies because it accesses the brain freely; it has also been used for smoking cessation in humans. It preferentially inhibits α3- and then α4-type nAChR.[6] Its mechanism of action is not understood.
- *Chlorisondamine* is a ganglionic blocker that produces a remarkably long lasting (>2 weeks) blockade of nAChR.[32]
- The N-methyl-D-aspartate (NMDA) channel blockers *MK 801* and *PCP* will inhibit nicotinic responses,[33] albeit at higher concentrations than for NMDA receptor blockade.
- *Bupropion*, a dopamine uptake inhibitor recently marketed for smoking cessation, has been found to be a non-competitive inhibitor of nAChR responses.[34]
- *Agonist molecules* (including nicotine and acetylcholine) at high concentration will block the nicotinic channel,[35] but it is unlikely that this will have functional significance *in vivo*.

Conclusions

In summary, nAChR are heterogeneous ligand gated ion channels, but in almost all cases the precise subunit composition of any native nAChR has not been unambiguously assigned. nAChR expressed artificially in cell lines or *Xenopus* oocytes may not therefore give a true reflection of the properties of nAChR existing in the brain.[36] The propensity of nAChR to desensitise will confound the interpretation of the effects of nicotine *in vivo*. Binding assays with radiolabelled or competing agonists will reflect the desensitised state of the nAChR, and the ability to bind to the receptor does not necessarily reflect the functional properties of the ligand. New drugs and toxins with nAChR subtype selectivity are emerging, and will help to unravel the contributions of nAChR subtypes to normal brain function and nicotine dependence.

2.2 Pharmacology and pharmacokinetics of nicotine

Chemistry of nicotine in tobacco smoke

Cigarette smoke is composed of volatile and particulate phases. Some 500 gaseous compounds including nitrogen, carbon monoxide (CO), carbon dioxide, ammonia, hydrogen cyanide and benzene, have been identified in the volatile phase which accounts for about 95% of the weight of cigarette smoke; the other 5% is accounted for by particulates. There are about 3,500 different compounds in the particulate phase, of which the major one is the alkaloid nicotine. Other alkaloids include nornicotine, anatabine, and anabasine.[37] The particulate matter without its alkaloid and water content is called tar. Many carcinogens, including polynuclear aromatic hydrocarbons, N-nitrosamines and aromatic amines, have been identified in cigarette tar.

Nicotine is a tertiary amine consisting of a pyridine and a pyrrolidine ring. There are two stereoisomers of nicotine: *(S)-nicotine* is the active isomer which binds to nicotinic cholinergic receptors and is found in tobacco. During smoking, some racemisation takes place, and small quantities of *(R)-nicotine*, a weak agonist of cholinergic receptors, are found in cigarette smoke.

Absorption of nicotine from tobacco products

Nicotine is distilled from burning tobacco, and small droplets of tar containing nicotine are inhaled and deposited in the small airways and alveoli. Nicotine is a weak base, and thus its absorption across cell membranes depends on the pH. The pH of smoke from most American cigarettes (blonde tobacco) is acidic (pH 5.5). At this pH, nicotine is mostly ionised and does not freely cross cell membranes. Consequently, nicotine from the blonde tobacco cigarette smoke is not absorbed through the buccal mucosa. However, the pH of smoke from tobacco in pipes and cigars is alkaline (pH 8.5), at which pH nicotine is mostly unionised and well absorbed from the mouth.

When nicotine from cigarette smoke reaches the small airways and the alveoli of the lung, it is buffered to physiological pH and rapidly absorbed into the pulmonary alveolar capillary and venous circulation, and hence directly into systemic arterial blood. From here, nicotine is distributed quickly throughout the body. It takes about 10–19 seconds for nicotine to reach the brain. The arterial blood perfusing the brain contains levels of nicotine following cigarette smoking which exceed venous levels by a factor of two- to sixfold.[38,39] Levels of nicotine in the plasma as well as in the brain decline rapidly as a result of distribution to peripheral tissues, and of excretion and elimination. Since no current nicotine replacement therapy (NRT) formulation uses the pulmonary route of absorption, none can mimic either the extremely high and rapidly acquired arterial nicotine concentrations which occur when tobacco products are inhaled, or the rapid pharmacological effect that this produces. The typical time course of the increase in nicotine levels in venous blood after smoking a cigarette is also faster than after most nicotine replacement products (Fig 2.2)[40] (see Chapter 7.2 for further discussion of NRT).

When smokers smoke multiple cigarettes during the day, there are oscillations between peak and trough plasma nicotine levels. However,

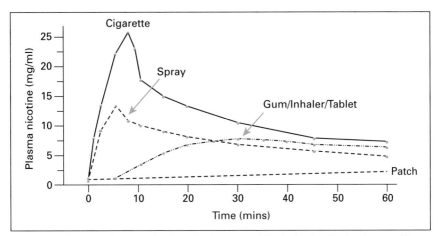

Fig 2.2. *Schematic diagram showing rise in venous blood nicotine levels after smoking a cigarette and after using different nicotine replacement therapy products, following overnight abstinence from cigarettes* (data adapted from Ref 40).

because of its half-life of two hours, nicotine accumulates over 6–8 hours, reaching levels in the plasma typically ranging from 20–40 ng/ml, which then fall progressively during the night.[41] There is considerable variation between people both in their plasma nicotine levels and in their intake of nicotine from a cigarette.[42,43] The smoker can manipulate the intake of nicotine from different cigarettes to achieve and maintain the desired level of nicotine (see Chapter 6) by changing puff volume, the number of puffs per cigarette, the intensity of puffing, the depth of inhalation, and by blocking ventilation holes in the filter.[44] No other nicotine product provides this degree of control over the rate and quantity of nicotine absorption.

Nicotine from chewing tobacco and snuff is absorbed through the oral and/or nasal mucosa. Plasma nicotine concentrations rise more slowly with these products than with cigarettes, reaching plateau levels by about 30 minutes, declining slowly over approximately two hours. Nicotine is continually released throughout the time of exposure.

The rapid absorption of nicotine from cigarette smoking, and the high arterial levels which reach the brain as a result, allow for rapid behavioural reinforcement from smoking. The falling nicotine levels between the smoking of individual cigarettes allow time for the brain nicotinic receptors to become somewhat resensitised between the cigarettes. Rapid delivery of nicotine to the brain also allows the smoker to manipulate and titrate the dose of nicotine from a cigarette to achieve a desired effect. Tolerance to the toxic effects of nicotine such as nausea rapidly develops and persists, while the reinforcing effects of nicotine are renewed with each cigarette. Thus, what is typically a noxious pharmacological experience for the novice smoker becomes an addictive pharmacological experience for the experienced smoker.

Nicotine metabolism

Nicotine is extensively metabolised, primarily in the liver but also to a small extent in the lung and brain. About 70–80% of nicotine is metabolised to cotinine via C-oxidation, and another 4% to nicotine N'-oxide.[45,46] There is considerable interindividual variability in the rate of metabolism of nicotine to cotinine,[43] but it is also established that regular smokers metabolise nicotine more slowly than non-smokers.[47,48] Several cytochrome P450 enzymes, as well as flavin mono-oxygenases, are purported to play a role in nicotine metabolism, but CYP2A6 appears to be the principal enzyme involved in converting nicotine to cotinine, via an intermediary metabolite nicotine $\Delta1(5')$-iminium ion.[49,50] Cotinine is further metabolised to trans-3'-hydroxycotinine, the major nicotine metabolite found in urine.[51] CYP2A6 is also the enzyme thought to be responsible for the oxidation of cotinine.[52] Cotinine has a much longer half-life than nicotine (14–20 hours), and consequently is used as a marker of nicotine intake.[53-55] Nicotine, cotinine, and trans-3'-hydroxycotinine are further metabolised by glucuronidation.[45] When using tobacco or nicotine medications, the strongest predictor of an individual's plasma nicotine levels is nicotine clearance. Cotinine levels are most strongly correlated with nicotine dose and, to a lesser extent, fractional conversion of nicotine to cotinine and cotinine clearance.[42] Renal clearance of nicotine depends on urine pH, being higher in acidic urine and lower in alkaline urine, and accounts for 2–35% of total nicotine clearance.[56]

Ethnic differences in nicotine metabolism have recently been demonstrated. African-Americans have been shown in several studies to have higher levels of cotinine, normalised for cigarettes smoked per day.[57,58] Recently, Perez-Stable et al administered deuterium-labelled nicotine and cotinine to African-American and Caucasian smokers, and found that the former metabolised cotinine more slowly than Caucasians.[59] African-Americans appear to do this both by slower oxidation to trans-3'-hydroxycotinine and by slower N-glucuronidation. They were also shown to take in 20% more nicotine per cigarette, which means an intake of 20% more tobacco smoke per cigarette. This may be related to the fact that the majority of African-Americans smoke mentholated cigarettes, whereas relatively few Caucasians smoke such cigarettes. Menthol cools the airways and might be associated with a greater volume or depth of inhalation. Taking a greater dose of nicotine per cigarette and also metabolising cotinine more slowly explain the higher cotinine levels per cigarette in African-Americans than in Caucasians. Greater smoke intake per cigarette might also explain the higher lung cancer risks, normalised for cigarette consumption, observed in African-Americans.[60]

Based on the idea that smokers regulate levels of nicotine in their bodies by adjusting how many cigarettes they smoke or how they smoke them, it is reasonable to speculate that smokers who metabolise nicotine more rapidly may need to take in more cigarette smoke, and vice versa. Pursuing this idea with respect to nicotine addiction, Pianezza *et al*[61] have studied the prevalence of mutant genes for CYP2A6, the liver enzyme primarily responsible for nicotine metabolism. They have reported that the presence of a CYP2A6*v1 mutant allele, presumably reflecting slower than normal nicotine metabolism, is associated with a lower risk of progression from experimental to addictive smoking. This study was in a relatively small group of smokers, and the effect of the CYP2A6*v1 alleles on the rate of nicotine metabolism has not yet been demonstrated, so the findings remain speculative.

Cardiovascular, endocrine and metabolic effects of nicotine

Nicotine effects on the cardiovascular system are mediated by sympathetic neural stimulation associated with an increase in the levels of circulating catecholamines. Nicotine causes sympathetic stimulation through central and peripheral mechanisms. Central nervous system (CNS)-mediated mechanisms include activation of peripheral chemoreceptors, particularly the carotid chemoreceptor, and direct effects on the brain stem and spinal cord.[62] Peripheral mechanisms include release of catecholamines from the adrenal glands and vascular nerve endings. These effects of nicotine result in an acute increase in heart rate and blood pressure when nicotine is delivered via cigarette smoking, chewing gum, nasal spray or intravenous (IV) infusion.[43,63,64] Transdermal nicotine causes less intense changes.[65] Substantial, but incomplete, tolerance develops to the cardiovascular effects of nicotine, with a short half-life of approximately 35 minutes; there is a persistent effect of nicotine, which is about 20% of the predicted effect if tolerance did not exist.[66] After brief dosing, there is no acute development of tolerance when the arterial plasma nicotine levels are determined. The difference in the observations between venous and arterial plasma levels may be because the levels of nicotine in arterial blood reflect the concentration of nicotine at the receptors, whereas the concentrations of nicotine in the venous blood reflect the levels of nicotine after distribution to the tissues. There is a lag time between the decline in venous compared to arterial levels, which could account for 'pseudotolerance', as assessed by venous blood level-response curves.[38,67]

Nicotine differentially affects blood flow to different organs, causing vasoconstriction in some vascular beds (eg skin) and vasodilatation in others (eg skeletal muscle). Cutaneous vasoconstriction results in a decrease in the fingertip temperature.[43] Nicotine induces vasoconstriction in coronary arteries, as evidenced by a lack of increased blood flow in response to increased oxygen demand,[68] and by direct observation, particularly in atherosclerotic arteries.[69–71] Coronary vasoconstriction appears to be mediated by catecholamines, and can be abolished by the α-adrenergic blocker phentolamine.[70]

Nicotine affects the metabolic rate, and smokers weigh on average 4 kg less than non-smokers.[72] The lower weight is maintained by an increase in metabolic rate, with concomitant appetite suppression.[72] Both cigarette smoking and IV nicotine increase the metabolic rate.[73] Quitting cigarette smoking is associated with an increase in appetite and caloric intake, with a subsequent weight gain over 6–12 months. Thereafter, both caloric intake and weight return to baseline.[74]

Nicotine has a variety of endocrine effects, including release of ACTH and cortisol,[75] and of β-endorphin,[76] and has been shown to have analgesic effects.[77]

2.3 Pathophysiological effects and toxicity of nicotine

Nicotine has a number of toxic or adverse effects, some of which are also potentially relevant in disease pathogenesis. For the purposes of this discussion, these are categorised as:

* acute systemic effects
* local toxic effects
* chronic systemic toxicity.

Acute systemic effects

Acute systemic toxic effects of nicotine include:[78]

* CNS effects: headache, dizziness, insomnia, abnormal dreams, nervousness
* gastrointestinal (GI) distress: dry mouth, nausea, vomiting, dyspepsia, diarrhoea
* musculoskeletal symptoms: arthralgias, myalgia.

In general, in relation to NRT for smokers, these effects tend to be mild. Interpretation of the CNS effects of nicotine in smokers who have recently quit smoking is complicated by the potential emergence of nicotine withdrawal symptoms that can be similar to some of the toxic effects of nicotine.

Local toxic effects

Local toxic effects of nicotine include:

* sore mouth and mouth ulcers from nicotine gum
* cutaneous sweating, itching, burning, erythema from patch application
* nasal irritation with burning, itching, sneezing
* watery eyes with nicotine nasal spray.

The mechanism of these effects is complex, but it appears to include activation of local afferent neurones and axon reflexes, with release of vasodilators such as bradykinin, substance P and histamine.[79] Local reactions from skin patches generally resolve within 24–48 hours. Nasal irritation with the use of nicotine nasal spray usually resolves with the development of tolerance over 2–3 days.

Chronic systemic toxicity

The main areas of concern over chronic systemic toxic effects of nicotine relate to effects on:

* cardiovascular disease, particularly coronary heart disease and stroke
* the aggravation of hypertension
* delayed wound healing
* peptic ulcer disease
* pregnancy (discussed further in Section 2.4) and reproductive health.

Cardiovascular disease. As described in Section 2.2, nicotine exerts cardiovascular effects primarily by activating the sympathetic nervous system, resulting in an increase in heart rate, blood pressure and cardiac contractility, thereby increasing myocardial oxygen consumption and demand for blood flow. Nicotine may also limit coronary blood flow by constricting coronary arteries, an effect more prominent in individuals with underlying coronary atherosclerosis.[80] Nicotine has also been associated with coronary spasm.[81] Other important cardiovascular toxins in cigarette smoke include CO, which reduces oxygen delivery to the heart, and oxidant gases, which may be responsible for endothelial dysfunction and platelet activation. Effects on endothelial function and platelets, mediated by oxidant gases, may be responsible for the thrombosis and/or coronary vasoconstriction that further restricts blood flow to the heart.[82]

Nicotine *per se*, at least when administered transdermally, does not activate platelets and probably does not contribute to thrombosis.[83] Prostacyclin is an endothelial-derived vasodilator and inhibitor of platelet aggregation. Nicotine has been shown to inhibit prostacyclin synthesis *in vitro*,[84] but studies of smokers and of smokeless tobacco and nicotine patch users found no evidence of decreased prostacyclin production.[85,86]

Alterations in the lipid profile, with an increase in very low-density lipoprotein (VLDL) and low-density lipoprotein and a decrease in high-density lipoprotein (HDL) cholesterol, are believed to be important mechanisms in smoking-induced atherosclerosis. Nicotine, via release of catecholamines, increases lipolysis and releases free fatty acids which are then taken up by the liver.[87] This might be expected to promote the synthesis of VLDL and decrease the synthesis of HDL, consistent with the changes seen in smokers,[88] but such abnormalities have not been found in people undergoing NRT.[89,90]

Studies of the effects of smokeless tobacco provide evidence regarding the cardiovascular safety of nicotine. Smokeless tobacco users are exposed to the same levels of nicotine in the body as cigarette smokers, but are not exposed to tar, CO and oxidant gases.[91] There is a significant pharmacokinetic difference in the rate of absorption, with cigarette smoking producing much higher transient arterial blood concentrations than smokeless tobacco; this must be kept in mind as a caveat in comparing nicotine exposures from the two routes.

Snuff use results in acute cardiovascular effects similar to those with cigarette smoking: that is, an increase in heart rate and blood pressure.[92] Cigarette smoking has been shown to affect platelet activation, as evidenced by increased thromboxane (TX) A2 metabolite excretion, and to impair endothelial function, primarily by reducing the release of nitric oxide (NO), which has antiplatelet activity and is a vasodilator.[85,86,93,94] None of the effects on TXA2 or NO is seen with snuff users, suggesting that the effects of smoking on these physiological functions are not mediated by nicotine.[95] However, results from epidemiological studies of cardiovascular disease in snuff users are conflicting. One case-control study found no increased risk of myocardial infarction (MI) in snuff users,[96] whereas another cohort study reported an increased risk.[97] The reason for the discrepancies between these two studies is unclear.

Clinical trials of nicotine medication in patients with coronary artery disease provide another important source of information. Two controlled clinical trials of transdermal nicotine to aid smoking cessation in patients with cardiovascular disease have found no evidence that nicotine is injurious.[98,99] Importantly, many of the subjects in these studies continued to smoke while using transdermal nicotine, resulting in plasma nicotine levels that might have been higher than those seen with smoking alone. A study by Mahmarian *et al*[100] examined quantitative thallium myocardial perfusion defect size in smokers with coronary heart disease prescribed nicotine patches to aid smoking cessation. When these subjects used 21-mg nicotine patches, their blood nicotine and cotinine levels were twice those seen with smoking alone, but their expired CO levels were reduced by about 50% because they smoked fewer cigarettes. The total and reversible thallium perfusion defect sizes were significantly reduced during the patch use, despite the high nicotine levels. This study suggests that components of cigarette smoke other than nicotine are responsible for acute ischaemia. Finally, the large Lung Health Study in patients with chronic obstructive pulmonary disease (COPD) found no increase in cardiovascular disease in smokers using nicotine gum for as long as five years.[101] Thus, the clinical trial data to date support the idea that nicotine medication is not a significant risk factor for cardiovascular events, even in patients with coronary heart disease.

Sporadic case reports have been published describing patients with acute cardiovascular events during the use of NRT. They include descriptions of patients who developed atrial fibrillation, acute MI or stroke.[102-104] Some of these patients were smoking at the same time as they were using transdermal nicotine. There was no consistent pattern with respect to how long these individuals had been using nicotine, time of day of the adverse event, or any other factor clearly identifying these events as related to the pharmacological effects of NRT. It must be recognised that cardiovascular disease is common in the age group of smokers undergoing smoking cessation therapy, and that some adverse cardiovascular events are expected to occur by chance in any 1–3 month period (the duration of most courses of NRT). A US Food and Drug Administration advisory committee reviewed the cases of MI in people using nicotine patches in 1992 and judged the events not to be causally related to their use.

Aggravation of hypertension. Although cigarette smoking and nicotine *per se* acutely increase blood pressure, cigarette smoking is not a risk factor for chronic hypertension.[105] Conceivably, factors such as lower body weight or differences in dietary intake in smokers might confound any blood pressure elevation due to nicotine.

Progression of chronic hypertension to accelerated or malignant hypertension, however, is much more likely in cigarette smokers.[106,107] Nicotine may contribute to the acceleration of hypertension by aggravating vasoconstriction. Animal studies indicate that nicotine may reduce renal blood flow which, in a patient with marginal renal blood flow due to hypertensive vascular disease, could cause renal ischaemia and aggravate hypertension.[108] There is therefore concern about using nicotine therapy in patients with severe hypertension.

Delayed wound healing. A prominent cardiovascular effect of nicotine is to reduce skin blood flow and subcutaneous tissue oxygen.[109,110] Adequate blood flow to the skin is important in the process of wound healing. Animal and human studies indicate that exposure to cigarette smoke or nicotine impairs the healing of skin flaps after plastic surgery.[111-113] It is likely that nicotine exposure in people will also delay wound healing after surgical procedures, although few clinical data are available on this issue.

Acid peptic disease. There is evidence that cigarette smoking aggravates gastro-oesophageal reflux, an effect documented in humans using transdermal nicotine.[114,115] Cigarette smoking is a strong risk factor for the development, delayed healing and relapse of peptic ulcer disease.[116] Nicotine could contribute both to reflux disease and gastric ulcer by provoking reflux of bile,[115] and to duodenal ulceration by decreasing bicarbonate secretion, an effect that may be related to depression of prostaglandin synthesis.[117-119] However, the results of the Lung Health Study do not support a role for nicotine *per se* in causing peptic disease. This study found no evidence that nicotine gum used for several years increased the risk of peptic ulcer disease, but rather that gum use had a borderline protective effect.[101]

Cancer. Because of the strong causal link between tobacco use and cancer, there has been concern as to whether nicotine contributes to cancer aetiology. Nicotine has not been shown to be carcinogenic in animals. There are theoretical concerns about nicotine and cancer related to metabolic activation or to stimulation of nicotinic cholinergic receptors that regulate release of lung tumour growth factors.[120,121] Nicotine could also contribute to cancer if it is nitrosated to form carcinogenic tobacco-specific nitrosamines.[122] Tobacco-specific nitrosamines are found in tobacco itself, resulting from the reaction of nitrites and alkaloids in the cigarette tobacco curing process. Nitrosamines can also be formed in the GI tract after oral administration of secondary amines and nitrites. Human exposure to nitrites occurs in the diet, and nicotine enters the GI tract both by swallowing products such as nicotine from nicotine medications and also by diffusion of nicotine from the bloodstream and ionic trapping by the acidic gastric fluid. Studies of urinary concentrations of nicotine-derived nitrosamines in humans exposed to nicotine are underway. It is likely that some nitrosation occurs, but the unresolved question is whether the amount of nicotine-derived nitrosamines is sufficient to contribute to cancer.

Although there are some concerns, as noted above, the risk of nicotine-related cancer, if any, is likely to be small or insignificant in tobacco users who are exposed to high concentrations of many carcinogens.

2.4 Effects of nicotine on mother and fetus in pregnancy

An understanding of the effects of nicotine during pregnancy is important both in relation to how cigarette smoking produces its adverse effects and also in balancing the potential risks and benefits of NRT to aid smoking cessation in pregnant women. The injurious effects of smoking (and potentially nicotine) on pregnancy have been summarised in Section 1.4 and can be considered in two categories:

* those that affect the pregnancy itself
* those that specifically affect the fetus during and after pregnancy.

Pregnancy

It is still unclear whether the injurious effects of cigarette smoking in pregnancy are due to nicotine – and, if so, *which* effects. Numerous other toxins in cigarette smoke, such as CO, oxidant gases, and heavy metals including lead and cadmium, could adversely affect the placental circulation and/or fetal physiology and development. The best studied tobacco smoke toxins during pregnancy are nicotine and CO. CO binds tightly to maternal haemoglobin, and even more tightly to fetal haemoglobin, reducing oxygen carrying capacity and impairing oxygen release from blood to fetal tissues. The result of CO exposure in the range of 5–10% carboxyhaemoglobin (similar to that seen in smokers) is a significant reduction in oxygen delivery, with resulting fetal hypoxia.[123] Animal studies show that CO exposure during pregnancy can reduce birth weight,[124,125] produce functional and structural abnormalities in the fetal brain,[126,127] and result in cognitive behavioural abnormalities in the newborn.[128,129] Oxidant gases are believed to impair NO formation and release from endothelial cells, including those found in the placenta, and could contribute to placental vascular insufficiency.[94,130] Lead is a well studied reproductive toxin which, among other effects, produces cognitive impairment in the newborn.[131] Thus, any consideration of the risks and benefits of nicotine during pregnancy must consider the numerous other proven toxins that are part of tobacco smoke.

A major concern about the effects of nicotine *per se* during pregnancy is that nicotine could constrict placental blood vessels, producing a state of placental and fetal hypoperfusion. This phenomenon has been demonstrated in experimental animals receiving high doses of IV nicotine.[132-134] However, the human placental circulation has considerable circulatory reserve, and studies in pregnant smokers either smoking cigarettes or receiving nicotine *per se* have not demonstrated evidence of placental hypoperfusion.[135] Thus, the vascular insufficiency module of smoking related fetal injury can be questioned, and the role of nicotine in causing adverse obstetrical outcomes remains to be established.

The fetus

The other major concern is that nicotine may have direct adverse effects on the developing fetus. Experimental work by Slotkin[136] and others has shown that exposure of pregnant rodents to nicotine results in impaired development of nicotinic cholinergic and other brain receptors in the offspring. Offspring so exposed have also been shown to have behavioural abnormalities, and to cope less well with hypoxic stress – the latter being a putative model for sudden infant death syndrome. Until otherwise demonstrated, it must therefore be assumed that nicotine has potential injurious effects on the developing fetus. However, since cigarette smoking exposes the individual to both nicotine and many other toxins, it seems clear that smoking is likely to be far more hazardous than nicotine obtained from alternative, cleaner sources such as nicotine replacement products. The practical consequences of this for replacement therapy are discussed further in Chapter 7.

2.5 Animal self-administration and nicotine addiction

The concept of addiction has at its core the idea of compulsive use, as reflected in powerful drug-seeking and drug-taking behaviour. In IV self-administration (IVSA) experiments, animals learn to administer drugs to themselves. Typically, the animal has the opportunity to press a lever; when it does so, it receives an automatic IV infusion of a drug (through a chronically-indwelling venous catheter). Several animal species, notably rats and monkeys, will press levers to obtain injections of the 'classical' addictive drugs such as morphine, heroin, amphetamine, cocaine, barbiturates and benzodiazepines. Large amounts of these drugs can be self-administered in this way. Furthermore, animals will work very hard to obtain the drugs, for example, pressing a switch thousands of times, for hours on end, to obtain drugs. This drug-seeking and drug-taking behaviour can dominate the animals' behavioural repertoire to the detriment of normal behaviour, just as in cases of serious drug abuse in humans. In fact, under appropriate conditions, animals will administer to themselves most of the drugs abused by humans. IVSA in animals is therefore a suitable animal model for the study of drug dependence in humans.

It has been established that monkeys, dogs, rats and mice can all exhibit nicotine IVSA. Monkeys have pressed levers for nicotine at rates similar to those at which they pressed levers for cocaine.[137] In these experiments, the nicotine or cocaine was paired (associated) with brief flashes of light, which in this model were functionally equivalent to the smell and taste stimuli associated with smoke inhalation. Both the nicotine and the light stimuli served as rewards for these animals, the latter by virtue of association with the nicotine. Nicotine IVSA has also been demonstrated in monkeys using simpler procedures without associated light stimuli, although rates and consistency of responding for the drug were less striking.[138,139] Dogs have also learned to press pedals to activate IV injections of nicotine. Up to several hundred pedal presses were made to obtain a single injection of nicotine, indicating that its rewarding effect, although powerful, was less strong than that of cocaine.[140] Nicotine IVSA has also been reported in mice both with an 'acute' procedure in which stress may be a confounding factor[141,142] and in chronic IVSA experiments of the usual type.[7,143] There is also some evidence for IV self-administration of pure solutions of nicotine in human subjects,[144] but these studies do not seem to have been reported in full. The animals studied most extensively in nicotine self-administration experiments have been rats, and these results will be considered next.

Nicotine self-administration in rats

In 1989, Corrigall and Coen succeeded in developing a rat model for nicotine IVSA.[145] The rats learnt to press levers to obtain IV infusions of nicotine, but did not press an inactive (control) lever in the same test chamber. The rate of lever pressing was related to the dose of nicotine, and the lever pressing ceased if nicotine was no longer available. As in the experiments in monkeys and dogs mentioned above, the lever-pressing produced nicotine and no other substance. The nicotine served as a goal object (positive reinforcer) for these animals, much in the same way as other drugs of abuse and natural rewards.

These observations have been reproduced and extended in numerous published experiments from many different laboratories.[146–152] All these studies demonstrate that rats will self-administer solutions of pure nicotine in the absence of any other reward. The validity of the observation is supported by the finding that the plasma concentration of nicotine in rats during IVSA experiments can be close to that in heavy cigarette smokers who inhale.[153]

Some studies have failed to find robust nicotine IVSA. To understand these results, it is essential to recognise that the extent to which a drug is self-administered depends on a multitude of procedural, environmental and genetic factors. As discussed elsewhere,[154] it was at one time difficult to demonstrate even the self-administration of opiate drugs. Some studies have failed to show nicotine IVSA,[155] but these used a strain of rat shown subsequently to be particularly poor in performing this task;[148] strain differences in animal studies may reflect genetic factors that influence human use of tobacco. Nearly all successful studies in rats used rapid 'bolus' injections, and data suggest that rapid infusions support self-administration more effectively than slow infusions.[156] Other studies used relatively slow infusions of nicotine and obtained equivocal results.[157] It is, however, apparent in most experiments that nicotine is a weaker reinforcer than cocaine, its self-administration is acquired more slowly and maintained under a narrower range of conditions. It is unclear whether this reflects on either the appropriateness of the animal procedures to model the richness of the human environment or on the importance of other reinforcers in human tobacco use. Non-pharmacological sources of reinforcement may be significant,[158] and recent studies provide indirect support for the presence of other psychoactive substances in tobacco in addition to the nicotine.[159,160]

Summary

IVSA is the primary animal model for studying drug-taking behaviour. The species thought to self-administer nicotine (and other drugs of abuse) in this way include mice, monkeys, dogs and rats. The combination of broad cross-species similarity of these animal data, plus the exceptionally high validity of drug self-administration procedures, generally strongly suggest that they are a reliable guide to the human condition. (Additional detail may be found in several reviews.[154,161–164])

Studies have also shed light on the brain mechanisms underlying nicotine IVSA. Nicotinic receptors are found in many areas of mammalian brain, and their involvement in nicotine IVSA is supported by observations that the non-competitive nicotine antagonist mecamylamine can attenuate lever-pressing for nicotine. The competitive nicotinic antagonist DHβE blocks nicotine IVSA by an action on nicotinic receptors in the VTA of the mid-brain.[165] Nicotine acts in the VTA to activate the ascending mesolimbic dopamine system.[166,167] This neural pathway is also critically implicated in the reinforcing action of abused drugs such as amphetamine, cocaine and opioids. Like these substances, nicotine enhances the release of dopamine in some of the projection areas of the mesolimbic dopamine system, notably the nucleus accumbens.[167–171] All these drugs, including nicotine, produce dopamine release mainly in the shell area of the nucleus accumbens rather than in the core.[170,171] Dopamine antagonist drugs and selective neurotoxin-produced lesions of the dopamine-containing neurones of the nucleus accumbens strikingly attenuate nicotine IVSA.[172,173] Recent studies suggest an impairment of both dopamine release and nicotine IVSA in transgenic mice lacking the β2 subunit of the nicotinic receptor;[7] nicotinic receptors containing the α4β2 subtype are the most prevalent in mammalian brain, and these important observations suggest that they may be required for nicotine IVSA. The roles of nicotinic receptors containing α and other β subunits have yet to be evaluated in a similar manner. Overall, the results to date suggest that the ascending mesolimbic dopamine system is essential for nicotine IVSA. This is also the major known neuroanatomical and neurochemical mechanism of reward for the classical addictive drugs. Therefore, both the behavioural effects and the mechanisms of action of nicotine in the IVSA model resemble those for classical drugs of abuse such as heroin and cocaine.

2.6 Nicotine neurochemistry: nicotine receptor and brain reward systems

Addictive drugs exhibit two important characteristics:

1 They elicit effects within the brain which are pleasant or rewarding, and which reinforce self-administration of the drug in both experimental animals and human beings.
2 Following a period of chronic exposure, withdrawal of the drug may elicit an abstinence syndrome which an addict may also seek to avoid by continuing to take the drug.

The primary objective of this section is to present the evidence that nicotine exerts effects within the brain which may account both for its positive reinforcing properties and for the presence of an abstinence syndrome following cessation of exposure.

The neurobiology underlying the positive reinforcing properties of nicotine

Studies of the mechanisms underlying the positive reinforcing properties of addictive drugs have been significantly influenced by experiments with the psychostimulant drugs, amphetamine and cocaine. These have shown that the ability of these compounds to act as locomotor stimulants and to reinforce self-administration in experimental animals depends critically upon the fact that they enhance neurotransmission at dopamine synapses in the mesolimbic system of the brain.[174,175] This conclusion is supported most impressively by the fact that lesions of the mesolimbic pathway cause a marked attenuation of the locomotor stimulant properties of these drugs and their ability to serve as a reinforcer in a self-administration paradigm. Indeed, it has been suggested that the large increases in dopamine overflow in the nucleus accumbens evoked by cocaine and amphetamine result in such a powerful euphoriant effect that it alone accounts for the addiction to these drugs.[174]

The locomotor stimulant properties of nicotine and its ability to act as a reward in a self-administration paradigm also seem to depend upon the ability of nicotine to stimulate the dopamine-secreting neurones which innervate the principal terminal field of the mesolimbic system, the nucleus accumbens.[165,173,176] In recent years, the effects of nicotine on dopamine release from these neurones have been studied extensively using the technique of *in vivo* microdialysis, a procedure which can be used to investigate the effects of drugs on neurotransmitter release from discrete areas of the brain in conscious, freely moving animals. These studies have shown clearly that nicotine preferentially stimulates dopamine release from the neurones which project to the nucleus accumbens.[168] Mesolimbic dopamine neurones express the nicotinic receptors which are known to mediate the effects of nicotine within the brain.[177] These receptors are located both on the nerve terminal membranes in the nucleus accumbens and on the membranes of the dopamine-secreting neurones in the mid-brain which innervate the nucleus accumbens.[178] Although both groups of receptors may contribute to the effects of nicotine on dopamine release, there is convincing evidence that the responses to nicotine injections, given either IV or subcutaneously, are mediated at least predominantly by the receptors located on the cell bodies in the mid-brain.[166,179] This implies that the effects of nicotine on the system depend upon its ability to influence the flow of impulses to the terminal field. In this respect, nicotine differs from cocaine and amphetamine which exert their effects by binding to the presynaptic dopamine transporter located on the nerve terminal membranes.

Studies on the neurobiology of drug addiction must take account of the fact that addiction is a consequence of chronic or repeated exposure to the drug. It is important to understand, therefore, the ways in which brain responses are influenced by chronic exposure to the drug since these may be crucial to our understanding of the neural mechanisms underlying addiction. In studies with experimental animals, repeated administration of amphetamine or cocaine results in sensitisation to their effects on dopamine release in the nucleus accumbens, measured using *in vivo* microdialysis.[180] It has been suggested that this sensitisation may play a central role in the development of addiction, in particular that sensitisation of the pathway may facilitate the way in which behaviours associated with obtaining the drug are learned and with the process by which 'drug-liking' becomes 'drug-craving'.[181] It is significant, therefore, that, like amphetamine and cocaine, repetitive injections of nicotine can also result in sensitisation of its effects on dopamine release in the accumbens.[169] The mechanisms underlying sensitisation of the response remain to be established with certainty. However, they seem to involve costimulation of the NMDA receptor for glutamate, since both the development and expression of the sensitised dopamine response are attenuated or abolished by the administration of NMDA receptor antagonists.[182,183] Costimulation of NMDA receptors has also been implicated in the mechanisms underlying sensitisation to other psychostimulant drugs of abuse, and is probably associated with an increase in the burst firing of the neurones.[180,184] Thus, the effects of chronic nicotine on the pathway are similar in important respects to those of other psychostimulant drugs of addiction.

The conclusions concerning the role of mesoaccumbens dopamine pathways in nicotine addiction are almost entirely derived from studies with animal models. However, there is circumstantial evidence to suggest that the conclusions apply to the reinforcing effects of nicotine in tobacco smoke to the extent that the administration of a drug, haloperidol, which blocks the dopamine receptor in the brain, increases smoking in habitual smokers.[185] This is the anticipated response if nicotine reward depends upon increased dopamine release in the brain since it reflects an attempt to overcome the blockade produced by the antagonist.

Other neural responses to nicotine which may contribute to its positive effects on smoking

Many neurones in the brain express the neuronal nicotinic receptors at which nicotine acts and, as a result, the drug stimulates other pathways which may be important to the development of addiction. These pathways include the noradrenaline-secreting neurones of the locus coeruleus which project to the forebrain, many of the acetylcholine-secreting neurones found in the hippocampus and cortex and terminals which secrete the excitatory amino acid, glutamic acid, and the inhibitory amino acid, γ-aminobutyric acid.[177,186,187] The psychopharmacological consequences of the effects of nicotine on these neurones remain to be established. However, it seems likely that stimulation of the receptor located on glutamate-secreting terminals facilitates release of the transmitter,[186] and that stimulation of NMDA receptors located on the dopamine-secreting neurones in the VTA results in increased burst firing of the neurones, and thus an enhanced dopamine response to nicotine.[188,189] It also seems likely that the effects of nicotine on acetylcholine-secreting neurones may be implicated in the increase in arousal and attention sometimes associated with smoking.[190] In addition, the stimulatory effects on both acetylcholine and glutamate secretion in the hippocampus and cerebral cortex may mediate the improved cognitive function which has been reported for nicotine.[191] Improved vigilance, attention and cognition have all been cited by smokers as reasons why they smoke.

Evidence for desensitisation of neuronal nicotinic receptors

Regular smoking results in the accumulation of nicotine in blood during the 'smoking day'. The nicotine level subsequently falls during sleep as the drug is metabolised and cleared from the body.[192] Prolonged exposure to nicotine has been shown to cause desensitisation of many of the neuronal nicotinic receptors which mediate its effects in the brain. There is now good evidence that the plasma concentrations of nicotine commonly found in habitual smokers during the day are sufficient to desensitise the nicotinic receptors on the mesolimbic dopamine neurones which appear to mediate the rewarding properties of the drug which reinforce its self-administration.[193] As a result, the administration of a nicotine bolus no longer causes increased dopamine release in the nucleus accumbens.[194]

These results have significant consequences for the 'dopamine hypothesis' of nicotine addiction. They imply that many smokers may continue smoking under conditions in which nicotine is unlikely to stimulate the mesolimbic dopamine neurones, and that other neural mechanisms must probably also contribute to the 'rewarding' properties of the drug which reinforce addiction.

In this context, it is important to remember that tobacco smoking habits are heterogeneous and that people smoke cigarettes at varying frequencies and in different ways. The plasma nicotine concentration is likely to remain fairly stable through the day for people who smoke frequently, whereas for those who smoke less frequently, significant peaks and troughs of nicotine may be observed.[195] If the nicotine concentration in the trough falls below that required to desensitise the nicotinic receptors on mesolimbic dopamine neurones, each cigarette will be 'rewarded' with increased dopamine release; for these smokers, stimulation of dopamine release is probably the predominant mechanism underlying addiction to nicotine.

In contrast, it has been suggested that receptor desensitisation may be the response which is reinforced in frequent or heavy smokers.[191] For example, the nicotinic receptors located on noradrenaline-secreting neurones are desensitised by nicotine concentrations similar to those found in the plasma of many smokers.[196] This may contribute to the 'tranquillising' properties of tobacco smoke often reported by smokers exposed to environmental stressors.[191] It is important to remember that nicotine exerts its effects in the brain by acting at a family of nicotine receptors. Thus, it is possible that other neural responses, mediated by receptors more resistant to desensitisation, may also play an important role in nicotine addiction.

In experimental animals, chronic exposure to nicotine, using regimens which cause desensitisation of the catecholamine responses to nicotine, often causes an increase in the density of the nicotine receptors which bind nicotine with high affinity.[177] This increased density seems to reflect a decreased turnover of the receptor complex.[197,198] The psychopharmacological significance of this effect remains unclear, although it appears to be associated with repeated or prolonged exposure to concentrations of nicotine which cause desensitisation of the receptors. It seems unlikely, however, that the increased receptor density accounts for the sensitisation to nicotine discussed above because up-regulation of the receptors is not observed with dosing regimens which elicit the sensitised dopamine responses.[199] The increase in receptor density may nevertheless be significant to the mechanisms underlying nicotine addiction since they are also observed in brain tissue taken from humans who have been habitual smokers.[200]

Effects on brain 5-hydroxytryptamine pathways

In experimental animals, chronic nicotine treatment causes a regionally-selective reduction in the concentration of 5-hydroxytryptamine

(5-HT) in the hippocampus.[201] Subsequent studies have shown this to be associated with a reduction in the formation and release of 5-HT in this region of the brain, and that tolerance to this effect does not develop in animals treated chronically with nicotine.[202,203] The results suggest that chronic exposure to the drug causes repeated or prolonged reductions in the demand for 5-HT in the hippocampus by reducing the concentration and capacity to synthesise 5-HT in serotonergic terminals in the hippocampus. Other studies using human post-mortem tissue have shown that habitual smoking is also associated with a regionally-selective reduction in the concentration of 5-HT and its principal metabolite, 5-hydroxyindole acetic acid, in the hippocampus which is not observed in a majority of the other areas of the brain that have been studied (ie gyrus rectus, medulla oblongata, cerebellum).[204]

These data imply that the reduction in hippocampal 5-HT formation and release observed in experimental rats given nicotine also occurs in the hippocampus of habitual smokers. It therefore seems reasonable to suggest that the effect is mediated by the nicotine present in tobacco smoke. This conclusion is supported by the association of habitual smoking with an increase in the density of post-synaptic $5-HT_{1A}$ receptors in the hippocampus since upregulation of these receptors is known to occur following treatments which elicit a chronic reduction of 5-HT release in the hippocampus.[204]

The psychopharmacological consequences of the changes in hippocampal 5-HT elicited by nicotine remain to be established. Experimental studies have shown that increased stimulation of $5-HT_{1A}$ receptors in the hippocampus may be implicated in anxiety.[205] It is possible, therefore, that the decrease in 5-HT release evoked by nicotine could mediate the reductions in anxiety often consistently reported by many smokers. There is some experimental evidence that nicotine has anxiolytic properties in some tests,[206] although this is not a universal finding.[207] If this hypothesis is correct, when habitual smokers first quit the habit, 5-HT release in this area of the brain will no longer be suppressed and, as a result, increased feelings of anxiety, mediated by the increased density of postsynaptic $5-HT_{1A}$ receptors in the hippocampus, will be experienced. Thus, the increases in receptor density could also contribute to the symptoms often observed during the early stages of smoking cessation.[208]

Other reports suggest that the 5-HT projections to the hippocampus are involved more in adaptation to aversive stimuli, that impairments in the pathway contribute to the dysfunction of pituitary-adrenal activity seen in many patients who suffer from depression, and that these effects contribute to symptoms of the patients.[209,210] There is a growing body of evidence to suggest that occupation of 5-HT receptors, including specifically 5-HT$_{1A}$ receptors, in the hippocampus plays a pivotal role in the expression of the glucocorticoid receptors which exert an important inhibitory effect on pituitary-adrenal activity.[211] These receptors are thought to play a primary role in the mechanism by which we 'cope' with the stresses of daily life, especially those to which we are exposed repetitively. Experimental studies suggest that chronic treatment with nicotine inhibits the process underlying habituation of the adrenocortical response to an unavoidable stressor.[212] This observation is entirely consistent with the fact that habituation of this response seems to depend upon increased expression of receptors which respond to glucocorticoids in the hippocampus, and that this increased receptor expression is highly dependent upon increased 5-HT release in this area of the brain.[213,214]

It seems reasonable to suggest, therefore, first, that one important consequence of the effects of chronic nicotine on hippocampal 5-HT function is to attenuate the mechanism which mediates adaptation to environmental stress; secondly, that this may explain some of the symptoms, such as anxiety and depression, observed when habitual smokers stop smoking,[208] when they are no longer 'protected' from these symptoms by the effects of nicotine on other neural pathways within the limbic system of the brain which are also involved in responses to anxiogenic or stressful stimuli.

References

1 Changeux J-P, Edelstein SJ. Allosteric receptors after 30 years. *Neuron* 1998; **21**: 959–80.

2 McGehee DS, Role LW. Physiological diversity of nicotinic acetylcholine receptors expressed by vertebrate neurons. *Annu Rev Physiol* 1995; **57**: 521–46.

3 Conroy WG, Vernallis AB, Berg DK. The α5 gene product assembles with multiple acetylcholine receptor subunits to form distinctive receptor subtypes in brain. *Neuron* 1992; **9**: 679–91.

4 Luetje CW, Patrick J. Both α and β subunits contribute to the agonist sensitivity of neuronal nicotinic acetylcholine receptors. *J Neurol Sci* 1991; **11**: 837–45.

5 Lukas RJ, Changeux JP, Le Novere N, Albuquerque EX, *et al.* International Union of Pharmacology document. XX. Current status of the nomenclature for nicotinic acetylcholine receptors and their subunits. *Pharmacol Rev* 1999; **51**: 397–401.

6 Gotti C, Fornasari D, Clementi F. Human neuronal nicotinic receptors. *Prog Neurobiol* 1997; **53**: 199–237.

7 Picciotto MR, Zoli M, Rimondini R, Lena C, *et al.* β2-Subunit-containing acetylcholine receptors are involved in the reinforcing properties of nicotine. *Nature* 1998; **391**: 173–7.

8 Le Novere N, Zoli M, Changeux JP. Neuronal nicotinic receptor α6 subunit mRNA is selectively concentrated in catecholaminergic nuclei of the rat brain. *Eur J Neurosci* 1996; **8**: 2428–39.

9   Goldner FM, Dineley KT, Patrick JW. Immunohistochemical localisation of the nicotinic acetylcholine receptor subunit α6 to dopaminergic neurons in the substantia nigra and ventral tegmental area. *NeuroReport* 1997; **8**: 2739–42.

10 Elgoyhen AB, Johnson DS, Boulter J, Vetter DE, Heinemann S. Alpha9: an acetylcholine receptor with novel pharmacological properties expressed in rat cochlear hair cells. *Cell* 1994; **79**: 705–15.

11 Anand R, Peng X, Lindstrom J. Homomeric and native alpha7 acetylcholine receptors exhibit remarkably similar but non-identical pharmacological properties, suggesting that the native receptor is a heteromeric protein complex. *FEBS Lett* 1993; **327**: 241–6.

12 Yu CR, Role LW. Functional contribution of the alpha7 subunit to multiple subtypes of nicotinic receptors in embryonic chick sympathetic neurones. *J Physiol (Lond)* 1998; **509**: 651–65.

13 Cuevas J, Berg DK. Mammalian nicotinic receptors with α7 subunits that slowly desensitize and rapidly recover from αbungarotoxin blockade. *J Neurosci* 1998; **18**: 10335–44.

14 Gray R, Rajan AS, Radcliffe KA, Yakehiro M, Dani JA. Hippocampal synaptic transmission enhanced by low concentrations of nicotine. *Nature* 1996; **383**: 713–6.

15 Flores CM, Rogers SW, Pabreza LA, Wolfe BB, Kellar KJ. A subtype of nicotinic cholinergic receptor in rat brain is composed of α4 and β2 subunits and is upregulated by chronic nicotine treatment. *Mol Pharmacol* 1992; **41**: 31–7.

16 Marks MJ, Robinson SF, Collins AC. Nicotinic agonists differ in activation and desensitisation of [86]Rb+ efflux from mouse thalamic synaptosomes. *J Pharmacol Exp Ther* 1996; **277**: 1383–96.

17 Wonnacott S. Presynaptic nicotinic receptors. *Trends Neurosci* 1997; **20**: 92–8.

18 Picciotto MR, Zoli M, Lena C, Bessis A, *et al.* Abnormal avoidance learning in mice lacking functional high-affinity nicotine receptor in the brain. *Nature* 1995; **374**: 65–7.

19 Clarke PBS. The fall and rise of neuronal alpha-bungarotoxin binding proteins. *Trends Pharmacol Sci* 1992; **13**: 407–13.

20 Davies ARL, Hardick DJ, Blagbrough IS, Potter BVL, *et al.* Characterisation of the binding of [3H]-methyllycaconitine: a new radioligand for labelling alpha7-type neuronal acetylcholine receptors. *Neuropharmacology* 1999; **38**: 679–90.

21 Romano C, Goldstein A. Stereospecific nicotinic receptors on rat brain membranes. *Science* 1980; **210**: 647–50.

22 Nordberg A. Imaging of nicotinic receptors in human brain. In: Domino EF (ed). *Brain imaging of nicotine and tobacco smoking.* Ann Arbor, MI: NPP Books, 1995: 45–57.

23 Houghtling RA, Davila-Garcia MI, Kellar KJ. Characterisation of (±)-[3H]epibatidine binding to nicotinic cholinergic receptors in rat and human brain. *Mol Pharmacol* 1995; **48**: 280–7.

24 Zoli M, Lena C, Picciotto MR, Changeux J-P. Identification of four classes of brain nicotinic receptors using β2 mutant mice. *J Neurosci* 1998; **18**: 4461–72.

25 Schulz DW, Loring RH, Aizenman E, Zigmond RE. Autoradiographic localization of putative nicotinic receptors in the rat brain using ^{125}I-neuronal bungarotoxin. *J Neurosci* 1991; **11**: 287–97.

26 Conroy WG, Berg DK. Neurons can maintain multiple classes of nicotinic acetylcholine receptor distinguished by different subunit compositions. *J Biol Chem* 1995; **270**: 4424–31.

27   de Fiebre CM, Meyer EM, Henry JC, Muraskin SI, *et al.* Characterisation of a series of anabaseine-derived compounds reveals that the 3-(4)-dimethyl-aminocinnamylidine derivative is a selective agonist at neuronal nicotinic alpha7/125I-alpha-bungarotoxin receptor subtypes. *Mol Pharmacol* 1995; **47**: 164–71.

28   Alkondon M, Pereira EFR, Cortes WS, Maelicke A, Albuquerque EX. Choline is a selective agonist of alpha7 nicotinic acetylcholine receptors in the rat brain neurons. *Eur J Neurosci* 1997; **9**: 2734–42.

29   Holladay MW, Lebold SA, Nan-Horng L. Structure-activity relationships of nicotinic acetylcholine receptor agonists as potential treatments for dementia. *Drug Dev Res* 1995; **353**: 191–213.

30   Wonnacott S, Albuquerque EX, Bertrand D. Methyllycaconitine: a new probe that discriminates between nicotinic acetylcholine receptor subclasses. In: Conn M (ed). *Methods in neurosciences*, vol 12. New York: Academic Press, 1993: 263–75.

31   Olivera BM. Conus venom peptides, receptor and ion channel targets, and drug design: 50 million years of neuropharmacology. *Mol Cell Biol* 1997; **8**: 2101–9.

32   Clarke PBS, Chaudieu I, El-Bizri H, Boksa M, *et al.* The pharmacology of the nicotinic antagonist, chlorisondamine, investigated in rat brain and autonomic ganglion. *Br J Pharmacol* 1994; **111**: 397–405.

33   Ramoa AS, Alkondon M, Aracava Y, Irons J, *et al.* The anticonvulsant MK-801 interacts with peripheral and central nicotinic acetylcholine receptor ion channels. *J Pharmacol Exp Ther* 1990; **254**: 71–82.

34   Fryer JD, Lukas RJ. Noncompetitive functional inhibition at diverse, human nicotinic acetylcholine receptor subtypes by bupropion, phencyclidine, and Ibogaine. *J Pharmacol Exp Ther* 1999; **288**: 88–92.

35   Colquhoun D, Ogden D, Mathie A. Nicotinic acetylcholine receptors of nerve and muscle: functional aspects. *Trends Pharmacol Sci* 1987; **8**: 465–72.

36   Sivilotti LG, McNeil DK, Lewis TM, Nassar MA, *et al.* Recombinant nicotinic receptors expressed in *Xenopus* oocytes do not resemble native rat sympathetic ganglion receptors in single-channel behaviour. *J Physiol (Lond)* 1997; **500**: 123–38.

37   Hoffmann D, Djordjevic MV, Hoffmann I. The changing cigarette. *Prev Med* 1997; **26**: 427–34.

38   Gourlay SG, Benowitz NL. Arteriovenous differences in plasma concentration of nicotine and catecholamines and related cardiovascular effects after smoking, nicotine nasal spray, and intravenous nicotine. *Clin Pharmacol Ther* 1997; **62**: 453–63.

39   Henningfield JE, Stapleton JM, Benowitz NL, Grayson RF, London ED. Higher levels of nicotine in arterial than in venous blood after cigarette smoking. *Drug Alc Depend* 1993; **33**: 23–9.

40   Russell MAH. Nicotine intake and its regulation by smokers. In: Martin WR, Loon GRV, Iwamoto ET, Davis L (eds). *Tobacco smoking and nicotine: a neurobiological approach*. New York: Plenum Publishing Corporation, 1987: 25–50.

41   Benowitz NL, Kuyt F, Jacob P III. Circadian blood nicotine concentrations during cigarette smoking. *Clin Pharmacol Ther* 1982; **32**: 758–64.

42   Benowitz NL, Zevin S, Jacob P III. Sources of variability in nicotine and cotinine levels with use of nicotine nasal spray, transdermal nicotine, and cigarette smoking. *Br J Clin Pharmacol* 1997; **43**: 259–67.

43   Benowitz NL, Jacob P III, Jones RT, Rosenberg J. Interindividual variability in the metabolism and cardiovascular effects of nicotine in man. *J Pharmacol Exp Ther* 1982; **221**: 368–72.

44 Herning RI, Jones RT, Benowitz NL, Mines AH. How a cigarette is smoked determines nicotine blood levels. *Clin Pharmacol Ther* 1983; **33**: 84–90.

45 Benowitz NL, Jacob P III, Fong I, Gupta S. Nicotine metabolic profile in man: comparison of cigarette smoking and transdermal nicotine. *J Pharmacol Exp Ther* 1994; **268**: 296–303.

46 Benowitz NL, Jacob P III. Metabolism of nicotine to cotinine studied by a dual stable isotope method. *Clin Pharmacol Ther* 1994; **56**: 483–93.

47 Benowitz NL, Jacob P III. Nicotine and cotinine elimination pharmacokinetics in smokers and nonsmokers. *Clin Pharmacol Ther* 1993; **53**: 316–23.

48 Lee BL, Benowitz NL, Jacob P III. Influence of tobacco abstinence on the disposition kinetics and effects of nicotine. *Clin Pharmacol Ther* 1987; **41**: 474–9.

49 Cashman JR, Park SB, Yang ZC, Wrighton SA, *et al.* Metabolism of nicotine by human liver microsomes: stereoselective formation of trans-nicotine-N'-oxide. *Chem Res Toxicol* 1992; **5**: 639–46.

50 McCracken NW, Cholerton S, Idle JR. Cotinine formation by cDNA-expressed human cytochromes P450. *Med Sci Res* 1992; **20**: 877–8.

51 Neurath GB, Dunger M, Orth D, Pein FG. Trans-3'-hydroxycotinine as a main metabolite in urine of smokers. *Int Arch Occup Environ Health* 1987; **59**: 199–201.

52 Nakajima M, Yamamoto T, Nunoya K, Yokoi T, *et al.* Characterization of CYP2A6 involved in 3'-hydroxylation of cotinine in human liver microsomes. *J Pharmacol Exp Ther* 1996; **277**: 1010–5.

53 Zevin S, Jacob P III, Benowitz NL. Cotinine effects on nicotine metabolism. *Clin Pharmacol Ther* 1997; **61**: 649–54.

54 Benowitz NL, Kuyt F, Jacob P III, Jones RT, Osman AL. Cotinine disposition and effects. *Clin Pharmacol Ther* 1983; **34**: 139–42.

55 DeSchepper PJ, Van Hecken A, Van Rossum JM. Kinetics of cotinine after oral and intravenous administration to man. *Eur J Clin Pharmacol* 1987; **31**: 583–8.

56 Benowitz NL, Jacob P III. Nicotine renal excretion rate influences nicotine intake during cigarette smoking. *J Pharmacol Exp Ther* 1985; **234**: 153–5.

57 Wagenknecht LE, Cutter GR, Haley NJ, Sidney S, *et al.* Racial differences in serum cotinine levels among smokers in the Coronary Artery Risk Development in (Young) Adults Study. *Am J Pub Health* 1990; **80**: 1053–6.

58 Carabello RS, Giovino GA, Pechacek TF, Mowery PD, *et al.* Racial and ethnic differences in serum cotinine levels of cigarette smokers. *JAMA* 1998; **280**: 135–9.

59 Perez-Stable EJ, Herrera B, Jacob P III, Benowitz NL. Nicotine metabolism and intake in black and white smokers. *JAMA* 1998; **280**: 152–6.

60 Harris RE, Zang EA, Anderson JI, Wynder EL. Race and sex differences in lung cancer risk associated with cigarette smoking. *Int J Epidemiol* 1993; **22**: 592–9.

61 Pianezza ML, Sellers EM, Tyndale RF. Nicotine metabolism defect reduces smoking. *Nature* 1998; **393**: 750.

62 Su C. Actions of nicotine and smoking on circulation. *Pharmacol Ther* 1982; **17**: 129–41.

63 Benowitz NL, Porchet H, Sheiner L, Jacob P III. Nicotine absorption and cardiovascular effects with smokeless tobacco use: comparison with cigarettes and nicotine gum. *Clin Pharmacol Ther* 1988; **44**: 23–8.

64 Sutherland G, Russell MAH, Stapleton J, Feyerabend C, Ferno O. Nasal nicotine spray: a rapid nicotine delivery system. *Psychopharmacology* 1992; **108**: 512–8.

65 Benowitz NL, Fitzgerald GA, Wilson M, Zhang Q. Nicotine effects on eicosanoid formation and hemostatic function: comparison of transdermal nicotine and cigarette smoking. *J Am Coll Cardiol* 1993; **22**: 1159–67.

66 Porchet HC, Benowitz NL, Sheiner LB. Pharmacodynamic model of tolerance: application to nicotine. *J Pharmacol Exp Ther* 1988; **244**: 231–6.

67 Porchet HC, Benowitz NL, Sheiner LB, Copeland JR. Apparent tolerance to the acute effect of nicotine results in part from distribution kinetics. *J Clin Invest* 1987; **80**: 1466–71.

68 Moreyra AE, Lacy CR, Wilson AC, Kumar A, Kostis JB. Arterial blood nicotine concentration and coronary vasoconstrictive effect of low nicotine smoking. *Am Heart J* 1992; **124**: 392–7.

69 Kaijser L, Berglund B. Effect of nicotine on coronary blood-flow in man. *Clin Physiol* 1985; **5**: 541–52.

70 Winniford MD, Wheelan KR, Kremers MS, Ugolini V, *et al.* Smoking-induced coronary vasoconstriction in patients with atherosclerotic coronary artery disease: evidence for adrenergically mediated alterations in coronary artery tone. *Circulation* 1986; **73**: 662–7.

71 Quillen JE, Rossen JD, Oskarsson HJ, Minor RL Jr, *et al.* Acute effect of cigarette smoking on the coronary circulation: constriction of epicardial and resistance vessels. *J Am Coll Cardiol* 1993; **22**: 642–7.

72 Perkins KA. Metabolic effects of cigarette smoking. *J Appl Physiol* 1992; **72**: 401–9.

73 Arcavi L, Jacob P III, Hellerstein M, Benowitz NL. Divergent tolerance to metabolic and cardiovascular effects of nicotine in smokers with low and high levels of cigarette consumption. *Clin Pharmacol Ther* 1994; **56**: 55–64.

74 Perkins KA. Weight gain following smoking cessation. *J Consult Clin Psychol* 1993; **61**: 768–77.

75 Baron JA, Comi RJ, Cryns V, Brinck-Johnsen T, Mercer NG. The effect of cigarette smoking on adrenal cortical hormones. *J Pharmacol Exp Ther* 1995; **272**: 151–5.

76 Seyler LE, Pomerleau OF, Fertig JB, Hunt D, Parker K. Pituitary hormone response to cigarette smoking. *Pharmacol Biochem Behav* 1986; **24**: 159–62.

77 Pomerleau OF, Turk DC, Fertig JB. The effects of cigarette smoking on pain and anxiety. *Addict Behav* 1984; **9**: 265–71.

78 Palmer KJ, Buckley MM, Faulds D. Transdermal nicotine. A review of its pharmacodynamic and pharmacokinetic properties, and therapeutic efficacy as an aid to smoking cessation. *Drugs* 1992; **44**: 498–529.

79 Smith EW, Smith KA, Maibach HI, Andersson PO, *et al.* The local side effects of transdermally absorbed nicotine. *Skin Pharmacol* 1992; **5**: 69–76.

80 Kaijser L, Berglund B. Effect of nicotine on coronary blood-flow in man. *Clin Physiol* 1985; **5**: 541–52.

81 Maouad J, Fernandez F, Hebert JL, Zamani K, *et al.* Cigarette smoking during coronary angiography: diffuse or focal narrowing (spasm) of the coronary arteries in 13 patients with angina at rest and normal coronary angiograms. *Cathet Cardiovasc Diagn* 1986; **12**: 366–75.

82 Celermajer DS. Endothelial dysfunction: does it matter? Is it reversible? *J Am Coll Cardiol* 1997; **30**: 325–33.

83 Benowitz NL, Fitzgerald GA, Wilson M, Zhang Q. Nicotine effects on eicosanoid formation and hemostatic function: comparison of transdermal nicotine and cigarette smoking. *J Am Coll Cardiol* 1993; **22**: 1159–67.

84 Wennmalm A, Alster P. Nicotine inhibits vascular prostacyclin but not platelet thromboxane formation. *Gen Pharmacol* 1983; **14**: 189–91.

85 Nowak J, Murray JJ, Oates JA, FitzGerald GA. Biochemical evidence of a chronic abnormality in platelet and vascular function in healthy individuals who smoke cigarettes. *Circulation* 1987; **76**: 6–14.

86 Rüngemark C, Benthin G, Granstrom EF, Persson L, *et al.* Tobacco use and urinary excretion of thromboxane A2 and prostacyclin metabolites in women stratified by age. *Circulation* 1992; **86**: 1495–500.

87 Hellerstein MK, Benowitz NL, Neese RA, Schwartz J, *et al.* Effects of cigarette smoking and its cessation on lipid metabolism and energy expenditure in heavy smokers. *J Clin Invest* 1994; **93**: 265–72.

88 Craig WY, Palomaki GE, Haddow JE. Cigarette smoking and serum lipid and lipoprotein concentrations: an analysis of published data. *Br Med J* 1989; **298**: 784–8.

89 Quensel M, Agardh C-D, Nilsson-Ehle P. Nicotine does not affect plasma lipoprotein concentrations in healthy men. *Scand J Clin Lab Invest* 1989; **49**: 149–53.

90 Thomas GAO, Davies SV, Rhodes J, Russell MAH, *et al.* Is transdermal nicotine associated with cardiovascular risk? *J R Coll Physicians Lond* 1995; **29**: 392–6.

91 Benowitz NL, Jacob P III, Yu L. Daily use of smokeless tobacco: systemic effects. *Ann Intern Med* 1989; **111**: 112–6.

92 Benowitz NL, Porchet H, Sheiner L, Jacob P III. Nicotine absorption and cardiovascular effects with smokeless tobacco use: comparison with cigarettes and nicotine gum. *Clin Pharmacol Ther* 1988; **44**: 23–8.

93 Andersson K, Eneroth P, Fuxe K, Harfstrand A. Effects of acute intermittent exposure to cigarette smoke on hypothalamic and preoptic catecholamine nerve terminal systems and on neuroendocrine function in the diestrous rat. *Naunyn-Schmiedeberg's Arch Pharmacol* 1988; **337**: 131–9.

94 Celermajer DS, Sorensen KE, Georgakopoulos D, Bull C, *et al.* Cigarette smoking is associated with dose-related and potentially reversible impairment of endothelium-dependent dilation in healthy young adults. *Circulation* 1993; **88**: 2149–55.

95 Wennmalm A, Benthin G, Granstrom EF, Persson L, *et al.* Relation between tobacco use and urinary excretion of thromboxane A2 and prostacyclin metabolites in young men. *Circulation* 1991; **83**: 1698–704.

96 Huhtasaari F, Asplund K, Lundberg V, Stegmayr B, Wester PO. Tobacco and myocardial infarction: is snuff less dangerous than cigarettes? *Br Med J* 1992; **305**: 1252–6.

97 Bolinder G, Alfredsson L, Englund A, deFaire U. Smokeless tobacco use and increased cardiovascular mortality among Swedish construction workers. *Am J Pub Health* 1994; **84**: 399–404.

98 Joseph AM, Norman SM, Ferry LH, Prochazka AV, *et al.* The safety of transdermal nicotine as an aid to smoking cessation in patients with cardiac disease. *N Engl J Med* 1996; **335**: 1792–8.

99 Working Group for the Study of Transdermal Nicotine in Patients with Coronary Artery Disease. Nicotine replacement therapy for patients with coronary artery disease. *Arch Intern Med* 1994; **154**: 989–95.

100 Mahmarian JJ, Moye LA, Nasser GA, Nagueh SF, *et al.* Nicotine patch therapy in smoking cessation reduces the extent of exercise-induced myocardial ischemia. *J Am Coll Cardiol* 1997; **30**: 125–30.

101 Murray RP, Bailey WC, Daniels K, Bjornson WM, *et al.* Safety of nicotine polacrilex gum used by 3,094 participants in the Lung Health Study. *Chest* 1996; **109**: 438–45.

102 Dacosta A, Guy JM, Tardy B, Gonthier R, *et al.* Myocardial infarction and nicotine patch: a contributing or causative factor? *Eur Heart J* 1993; **14**: 1709–11.

103 Pierce JR. Stroke following application of a nicotine patch. *Ann Pharmacother* 1994; **28**: 402.

104 Stewart PM, Catterall JR. Chronic nicotine ingestion and atrial fibrillation. *Br Heart J* 1985; **54**: 222–3.

105 Green MS, Jucha E, Luz Y. Blood pressure in smokers and nonsmokers: epidemiologic findings. *Am Heart J* 1986; **111**: 932–40.

106 Isles C, Brown JJ, Cummings AM, Lever AF, *et al.* Excess smoking in malignant-phase hypertension. *Br Med J* 1979; **1**: 579–81.

107 Petitti DB, Klatsky AL. Malignant hypertension in women aged 15 to 44 years and its relation to cigarette smoking and oral contraceptives. *Am J Cardiol* 1983; **52**: 297–8.

108 Downey HF, Crystal GJ, Bashour FA. Regional renal and splanchnic blood flows during nicotine infusion: effects of beta adrenergic blockade. *J Pharmacol Exp Ther* 1981; **216**: 363–7.

109 Jensen JA, Goodson WH, Hopf HW, Hunt TK. Cigarette smoking decreases tissue oxygen. *Arch Surg* 1991; **126**: 1131–4.

110 Rao VK, Morrison WA, O'Brien BM. Effect of nicotine on blood flow and patency of experimental microvascular anastomosis. *Ann Plast Surg* 1983; **11**: 206–9.

111 Forrest CR, Pang CY, Lindsay WK. Pathogenesis of ischemic necrosis in random-pattern skin flaps induced by long-term low-dose nicotine treatment in the rat. *Plast Reconstr Surg* 1991; **87**: 518–28.

112 Lawrence WT, Murphy RC, Robson MC, Heggers JP. The detrimental effect of cigarette smoking on flap survival: an experimental study in the rat. *Br J Plast Surg* 1984; **37**: 216–9.

113 Rees TD, Liverett DM, Guy CL. The effect of cigarette smoking on skin-flap survival in the face lift patient. *Plast Reconstr Surg* 1984; **73**: 911–5.

114 Rahal S, Wright RA. Transdermal nicotine and gastro-oesophageal reflux. *Am J Gastroenterol* 1995; **90**: 919–21.

115 Rees SDW, Rhodes J. Bile reflux in gastroesophageal disease. *Clin Gastroenterol* 1977; **6**: 179–200.

116 Kikendall JW, Evaul J, Johnson LF. Effect of cigarette smoking on gastro-intestinal physiology and non-neoplastic digestive disease. *J Clin Gastroenterol* 1984; **6**: 65–78.

117 Endoh K, Leung FW. Effects of smoking and nicotine on the gastric mucosa: a review of clinical and experimental evidence. *Gastroenterology* 1994; **107**: 864–78.

118 Ganstam SO, Jonson C, Fandriks L, Holm L, Flemstron G. Effects of cigarette smoke and nicotine on duodenal bicarbonate secretion in the rabbit and the rat. *J Clin Gastroenterol* 1990; **12**(Suppl 1): S19–24.

119 Murthy SNS, Dinoso VP Jr, Clearfield HR, Chey WY. Simultaneous measurement of basal pancreatic, gastric acid secretion, plasma gastrin, and secretin during smoking. *Gastroenterology* 1977; **73**: 758–61.

120 Castagnoli N, Liu X, Shigenaga MK, Wardrop R, Castagnoli K. Studies on the metabolic fate of (S)-nicotine and its pyrrolic analog b-nicotyrine. In: Benowitz NL (ed). *Nicotine safety and toxicity.* New York: Oxford University Press, 1998: 57–65.

121 Schuller HM. Nicotine and lung cancer. In: Benowitz NL (ed). *Nicotine safety and toxicity.* New York: Oxford University Press, 1998: 77–85.

122 Hecht SS. Biochemistry, biology, and carcinogenicity of tobacco-specific N-nitrosoamines. *Chem Res Toxicol* 1998; **11**: 559–603.

123 Longo LD. The biological effects of carbon monoxide on the pregnant woman, fetus, and newborn infant. *Am J Obstet Gynecol* 1977; **129**: 69–103.

124 Garvey DJ, Longo LD. Chronic low level maternal carbon monoxide exposure and fetal growth and development. *Biol Reprod* 1978; **19**: 8–14.

125 Lynch A-M, Bruce NW. Placental growth in rats exposed to carbon monoxide at selected stages of pregnancy. *Biol Neonate* 1989; **56**: 151–7.

126 Fechter LD, Annau Z. Toxicity of mild prenatal carbon monoxide exposure. *Science* 1977; **197**: 680–2.

127 Storm JE, Valdes JJ, Fechter LD. Postnatal alterations in cerebellar GABA content, GABA uptake and morphology following exposure to carbon monoxide early in development. *Dev Neurosci* 1986; **8**: 251–61.

128 Mactutus CF, Fechter LD. Prenatal exposure to carbon monoxide: learning and memory deficits. *Science* 1984; **223**: 409–11.

129 Mactutus CF, Fechter LD. Moderate prenatal carbon monoxide exposure produces persistent, and apparently permanent, memory deficits in rats. *Teratology* 1985; **31**: 1–12.

130 Sooranna SR, Morris NH, Steer PJ. Placental nitric oxide metabolism. *Reprod Fertil Dev* 1995; **7**: 1525–31.

131 Bellinger D, Leviton A, Waternaux C, Needleman H, Rabinowitz M. Longitudinal analyses of prenatal and postnatal lead exposure and early cognitive development. *N Engl J Med* 1987; **316**: 1037–43.

132 Ayromlooi J, Desiderio D, Tobias M. Effect of nicotine sulfate on the hemodynamics and acid base balance of chronically instrumented pregnant sheep. *Dev Pharmacol Ther* 1981; **3**: 205–13.

133 Resnik R, Brink GW, Wilkes M. Catecholamine-mediated reduction in uterine blood flow after nicotine infusion in the pregnant ewe. *J Clin Invest* 1979; **63**: 1133–6.

134 Suzuki K, Horiguchi T, Comas-Urrutia AC, Mueller-Heubach E, *et al.* Pharmacologic effects of nicotine upon the fetus and mother in the rhesus monkey. *Am J Obstet Gynecol* 1971; **11**: 1092–101.

135 Lambers DS, Clark KE. The maternal and fetal physiologic effects of nicotine. *Semin Perinatol* 1996; **20**: 115–26.

136 Slotkin TA. Fetal nicotine or cocaine exposure: which one is worse? *J Pharmacol Exp Ther* 1998; **285**: 931–45.

137 Goldberg SR, Spealman RD, Goldberg DM. Persistent behavior at high rates maintained by intravenous self-administration of nicotine. *Science* 1981; **214**: 573–5.

138 Deneau GA, Inoki R. Nicotine self-administration in monkeys. *Ann NY Acad Sci* 1967; **142**: 277–9.

139 Spealman RD, Goldberg SR. Maintenance of schedule-controlled behavior by intravenous injections of nicotine in squirrel monkeys. *J Pharmacol Exp Ther* 1982; **223**: 402–8.

140 Risner ME, Goldberg SR. A comparison of nicotine and cocaine self-administration in the dog: fixed-ratio and progressive-ratio schedules of intravenous drug infusion. *J Pharmacol Exp Ther* 1983; **224**: 319–26.

141 Martellotta MC, Kuzmin A, Zvartau E, Cossu G, *et al.* Isradipine inhibits nicotine intravenous self-administration in drug-naive mice. *Pharmacol Biochem Behav* 1995; **52**: 271–4.

142 Rasmussen T, Swedberg MDB. Reinforcing effects of nicotinic compounds: intravenous self-administration in drug-naive mice. *Pharmacol Biochem Behav* 1998; **60**: 567–73.

143 Stolerman IP, Naylor C, Elmer GI, Goldberg SR. Discrimination and self-administration of nicotine by inbred strains of mice. *Psychopharmacology* 1999; **141**: 297–306.

144 Henningfield JE, Goldberg SR. Nicotine as a reinforcer in human subjects and laboratory animals. *Pharmacol Biochem Behav* 1983; **19**: 989–92.

145 Corrigall WA, Coen KM. Nicotine maintains self-administration in rats on a limited-access schedule. *Psychopharmacology* 1989; **99**: 473–8.

146 Tessari M, Valerio E, Chiamulera C, Beardsley PM. Nicotine reinforcement in rats with histories of cocaine self-administration. *Psychopharmacology* 1995; **121**: 282–3.

147 Donny EC, Caggiula AR, Knopf S, Brown C. Nicotine self-administration in rats. *Psychopharmacology* 1995; **122**: 390–4.

148 Donny EC, Caggiula AR, Mielka MM, Jacobs KS, *et al.* Acquisition of nicotine self-administration in rats: the effects of dose, feeding schedule, and drug contingency. *Psychopharmacology* 1998; **136**: 83–90.

149 Epping-Jordan MP, Markou A, Koob GF. Intravenous nicotine self-administration in rats. *Behav Pharmacol* 1996; **7**(Suppl 1): 35 (abstract).

150 Shoaib M, Schindler CW, Goldberg SR. Nicotine self-administration in rats: strain and nicotine pre-exposure effects on acquisition. *Psychopharmacology* 1997; **129**: 35–43.

151 Bardo MT, Green TA, Valone JM, Crooks PA, Dwoskin LP. Nornicotine is self-administered intravenously and reduces nicotine self-administration in rats. *Soc Neurosci Abstr* 1998; **24**: 752.

152 Valentine JD, Hokanson JS, Matta SG, Sharp BM. Self-administration in rats allowed unlimited access to nicotine. *Psychopharmacology* 1997; **133**: 300–4.

153 Shoaib M, Stolerman IP. Plasma nicotine and cotinine levels following intravenous nicotine self-administration in rats. *Psychopharmacology* 1999; **143**: 318–21.

154 Stolerman IP, Jarvis MJ. The scientific case that nicotine is addictive. *Psychopharmacology* 1995; **117**: 2–10.

155 Dworkin SI, Vrana SL, Broadbent J, Robinson JH. Comparing the reinforcing effects of nicotine, caffeine, methylphenidate and cocaine. *Med Chem Res* 1993; **2**: 593–602.

156 Wakasa Y, Takada K, Yanagita T. Reinforcing effect as a function of infusion speed in intravenous self-administration of nicotine in rhesus monkeys. *Nihon Shinkei Seishin Yakurigaku Zasshi* 1995; **15**: 53–9.

157 Pudiak CM, Bozarth MA. Assessing nicotine reinforcement with intravenous self-administration: tests using a standard cross-substitution procedure in cocaine-trained rats. *Addiction* 1997; **92**: 628.

158 Rose JE, Behm FM, Levin ED. Role of nicotine dose and sensory cues in the regulation of smoke intake. *Pharmacol Biochem Behav* 1993; **44**: 891–900.

159 Fowler JS, Volkow ND, Wang G-J, Pappas N, *et al.* Inhibition of monoamine oxidase B in the brains of smokers. *Nature* 1996; **379**: 733–6.

160 Fowler JS, Volkow ND, Wang GJ, Pappas N, *et al.* Brain monoamine oxidase A inhibition in cigarette smokers. *Proc Natl Acad Sci USA* 1996; **93**: 14065–9.

161 Stolerman IP. Psychopharmacology of nicotine: stimulus effects and receptor mechanisms. In: Iversen LL, Iversen SD, Snyder SH (eds). *Handbook of psychopharmacology*, vol. 19. New York: Plenum, 1987: 421–65.

162 Henningfield JE, Goldberg SR. Stimulus properties of nicotine in animals and human volunteers: a review. In: Seiden LS, Balster RL (eds). *Behavioral pharmacology: the current status*. New York: Alan R Liss, 1985: 433–49.

163 Rose JE, Corrigall WA. Nicotine self-administration in animals and humans: similarities and differences. *Psychopharmacology* 1997; **130**: 28–40.

164 Swedberg MDB, Henningfield JE, Goldberg SR. Nicotine dependency: animal studies. In: Wonnacott S, Russell MAH, Stolerman IP (eds). *Nicotine psychopharmacology: molecular, cellular and behavioural aspects*. Oxford: Oxford University Press, 1990: 38–76.

165 Corrigall WA, Coen KM, Adamson LK. Self-administered nicotine activates the mesolimbic dopamine system through the ventral tegmental area. *Brain Res* 1994; **653**: 278–84.

166 Nisell M, Nomikos GG, Svensson TH. Systemic nicotine-induced dopamine release in the rat nucleus accumbens is regulated by nicotinic receptors in the ventral tegmental area. *Synapse* 1994; **16**: 36–44.

167 Nisell M, Nomikos GG, Svensson TH. Infusion of nicotine in the ventral tegmental area or the nucleus accumbens of the rat differentially affects accumbal dopamine release. *Pharmacol Toxicol* 1994; **75**: 348–52.

168 Imperato A, Mulas A, Di Chiara G. Nicotine preferentially stimulates dopamine release in the limbic system of freely moving rats. *Eur J Pharmacol* 1986; **132**: 337–8.

169 Benwell MEM, Balfour DJK. The effects of acute and repeated nicotine treatment on nucleus accumbens dopamine and locomotor activity. *Br J Pharmacol* 1995; **105**: 849–56.

170 Pontieri FE, Tanda G, Orzi F, Di Chiara G. Effects of nicotine on the nucleus accumbens and similarity to those of addictive drugs. *Nature* 1996; **382**: 255–7.

171 Pontieri FE, Passarelli F, Calo L, Caronti B. Functional correlates of nicotine administration: similarity with drugs of abuse. *J Mol Med* 1998; **76**: 193–201.

172 Corrigall WA, Coen KM. Selective dopamine antagonists reduce nicotine self-administration. *Psychopharmacology* 1991; **104**: 171–6.

173 Corrigall WA, Franklin KBJ, Coen KM, Clarke PBS. The mesolimbic dopaminergic system is implicated in the reinforcing effects of nicotine. *Psychopharmacology* 1992; **107**: 285–9.

174 Wise RA, Bozarth MA. A psychostimulant theory of addiction. *Psychol Rev* 1987; **94**: 469–92.

175 Di Chiara G, Imperato A. Drugs abused by humans preferentially increase synaptic dopamine concentration in the mesolimbic system of freely moving rats. *Proc Natl Acad Sci USA* 1988; **85**: 5274–8.

176 Clarke PBS. Dopaminergic mechanisms in the locomotor stimulant effects of nicotine. *Biochem Pharmacol* 1990; **40**: 1427–32.

177 Wonnacott S. Characterization of brain nicotinic receptor sites. In: Wonnacott S, Russell MAH, Stolerman IP (eds). *Nicotine psychopharmacology: molecular, cellular and behavioural aspects.* Oxford: Oxford University Press, 1990: 226–77.

178 Clarke PBS, Pert A. Autoradiographic evidence for nicotinic receptors on nigrostriatal and mesolimbic dopaminergic neurones. *Brain Res* 1985; **348**: 355–8.

179 Benwell MEM, Balfour DJK, Lucchi HM. The influence of tetrodotoxin and calcium on the stimulation of mesolimbic dopamine activity evoked by systemic nicotine. *Psychopharmacology* 1993; **112**: 467–71.

180 Kalivas PW, Sorg BA, Hooks MS. The pharmacology and neural circuitry of sensitization to psychostimulants. *Behav Pharmacol* 1993; **4**: 315–34.

181 Robinson TE, Berridge KC. The neural basis of drug craving: an incentive-sensitisation theory of addiction. *Brain Res Rev* 1993; **18**: 374–418.

182 Balfour DJK, Birrell CE, Moran RJ, Benwell MEM. Effects of D-CPPene on mesoaccumbens dopamine responses to nicotine in the rat. *Eur J Pharmacol* 1996; **316**: 153–6.

183 Shoaib M, Benwell MEM, Akbar MT, Stolerman IP, Balfour DJK. Behavioural and neurochemical adaptations to nicotine in rats: influence of NMDA antagonists. *Br J Pharmacol* 1994; **111**: 1073–80.

184 Overton PG, Clark D. Burst firing of midbrain dopaminergic neurons. *Brain Res Rev* 1997; **25**: 312–34.

185 Dawe S, Geroda C, Russell MAH, Gray JA. Nicotine intake in smokers increases following a single dose of haloperidol. *Psychopharmacology* 1995; **117**: 110–5.

186 McGehee DS, Heath MJS, Gelber S, Devay P, Role LW Nicotine enhancement of fast excitatory synaptic transmission in CNS by presynatic receptors. *Science* 1995; **269**: 1692–6.

187 Lu Y, Grady S, Marks MJ, Picciotto M, *et al*. Pharmacological characterization of nicotinic receptor-stimulated GABA release from mouse brain synaptosomes. *J Pharmacol Exp Ther* 1998; **287**: 648–57.

188 Balfour DJK, Benwell MEM, Birrell CE, Kelly RJ, Al-Aloul M. Sensitization of the mesoaccumbens dopamine response to nicotine. *Pharmacol Biochem Behav* 1998; **59**: 1021–30.

189 Schilström B, Nomikos GG, Nisell M, Hertel P, Svensson TH. N-Methyl-D-aspartate receptor antagonism in the ventral tegmental area diminishes the systemic nicotine-induced dopamine release in the nucleus accumbens. *Neuroscience* 1998; **82**: 781–9.

190 Balfour DJK. The effects of nicotine on brain neurotransmitter systems. In: Balfour DJK (ed). *Nicotine and the tobacco smoking habit*. The International Encyclopaedia of Pharmacology and Therapeutics, Section 114. Oxford: Pergamon Press, 1984: 61–74.

191 Balfour DJK, Fagerström KO. Pharmacology of nicotine and its therapeutic use in smoking cessation and neurodegenerative disorders. *Pharmacol Therap* 1996; **72**: 51–81.

192 Benowitz NL, Jacob P III, Savanapridi C. Determinants of nicotine intake while chewing nicotine piracrilax gum. *Clin Pharmacol Therap* 1987; **41**: 467–73.

193 Pidoplichko VI, De Bias M, Williams JT, Dani JA. Nicotine activates and desensitises midbrain dopamine neurones. *Nature* 1997; **390**: 401–4.

194 Benwell MEM, Balfour DJK, Birrell CE. Desensitization of the nicotine-induced mesolimbic dopamine responses during constant infusion with nicotine. *Br J Pharmacol* 1995; **114**: 454–60.

195 Russell MAH. Nicotine intake and its control over smoking. In: Wonnacott S, Russell MAH, Stolerman IP (eds). *Nicotine psychopharmacology: molecular, cellular and behavioural aspects*. Oxford: Oxford University Press, 1990: 374–410.

196 Benwell MEM, Balfour DJK. Regional variation in the effects of nicotine on catecholamine overflow in the rat brain. *Eur J Pharmacol* 1997; **325**: 13–20.

197 Marks MJ, Pauly JR, Gross SD, Deneris ES, *et al*. Nicotine binding and nicotinic receptor subunit RNA after chronic nicotine treatment. *J Neurosci* 1992; **12**: 2765–84.

198 Peng X, Gerzanich V, Anand R, Whiting PJ, Lindstrom J. Nicotine-induced increase in neuronal nicotinic receptors results from a decrease in the rate of receptor turnover. *Mol Pharmacol* 1994; **46**: 523–60.

199 Benwell MEM, Balfour DJK. Nicotine binding to brain tissue from drug naive and nicotine-treated rats. *J Pharm Pharmacol* 1985; **37**: 405–9.

200 Benwell MEM, Balfour DJK, Anderson JM. Evidence that smoking increases the density of nicotine binding sites in human brain. *J Neurochem* 1988; **50**: 1243–7.

201 Benwell MEM, Balfour DJK. Effects of nicotine administration and its withdrawal on plasma corticosterone and brain 5-hydroxyindoles. *Psychopharmacology* 1979; **63**: 7–11.

202 Benwell MEM, Balfour DJK. The effects of nicotine administration on 5-HT uptake and biosynthesis in rat brain. *Eur J Pharmacol* 1982; **84**: 71-7.

203 Ridley DL, Balfour DJK. The influence of nicotine on 5-HT overflow in the dorsal hippocampus of the rat. *Br J Pharmacol* 1997; **122**: 301.

204 Benwell MEM, Balfour DJK, Anderson JM. Smoking-associated changes in serotonergic systems of discrete regions of human brain. *Psychopharmacology* 1990; **102**: 68–72.

205 Andrews N, Hogg S, Gonzalez LE, File SE. 5-HT$_{1A}$ receptors in the median raphe and dorsal hippocampus may mediate anxiolytic and anxiogenic behaviours respectively. *Eur J Pharmacol* 1994; **264**: 259–64.

206 Brioni JD, O'Neil AB, Kim DJB, Buckley MJ, *et al.* Anxiolytic-like effects of the novel cholinergic channel activator, ABT-418. *J Pharmacol Exp Therap* 1994; **271**: 352–61.

207 Balfour DJK, Graham CA, Vale AL. Studies on the possible role of brain 5-HT systems and adrenocortical activity in behavioural responses to nicotine and diazepam. *Psychopharmacology* 1986; **90**: 528–32.

208 Fagerström KO, Schneider NG. Measuring nicotine dependence: a review of the Fagerström Tolerance Questionnaire. *J Behav Med* 1989; **12**: 159–82.

209 Deakin JFW, Graeff FG. 5-HT and mechanisms of defence. *J Psychopharmacol* 1991; **5**: 305–15.

210 Graeff FG, Guimaraes FS, De Andrade TG, Deakin JF. Role of 5-HT in stress, anxiety and depression. *Pharmacol Biochem Behav* 1996; **54**: 129–41.

211 Seckl JR, Fink G. Use of in situ hybridization to investigate the regulation of hippocampal corticosterone receptors by monoamines. *J Steroid Biochem Mol Biol* 1991; **40**: 685–8.

212 Benwell MEM, Balfour DJK. Effects of chronic nicotine administration on the response and adaptation to stress. *Psychopharmacology (Berlin)* 1982; **76**: 160–2.

213 Meany MJ, Aitken DH, Viau V, Sharma S, Sarrieau A. Neonatal handling alters the adrenocortical negative feedback sensitivity and hippocampal type-II glucocorticoid receptor-binding in the rat. *Neuroendocrinology* 1989; **50**: 597–604.

214 Yau JL, Kelly PA, Seckl JR. Increased glucocorticoid gene expression in the rat hippocampus following combined serotonergic and medial septal cholinergic lesions. *Mol Brain Res* 1994; **27**: 174–8.

3 | Psychological effects of nicotine and smoking in man

3.1 The effects of nicotine and smoking on mood and cognition
3.2 The nicotine withdrawal syndrome
3.3 Psychological dependence on nicotine and smoking

3.1 The effects of nicotine and smoking on mood and cognition

Demonstration of nicotine addiction and evidence of putative under-lying neurobiological mechanisms in animals do not establish conclusively that nicotine has psychological effects in man. Furthermore, much of the evidence on the psychological effects of nicotine in man is derived from studies of smokers, and the question therefore arises as to how much any psychological effect of smoking is in fact attributable to nicotine. Evidence relating to this question is reviewed in this chapter.

Smokers' perceptions of the psychological effects of smoking

From early on in their smoking career, smokers perceive that smoking provides certain psychological benefits. 'Feeling calmer' is typically the most commonly reported subjective effect of smoking, with 64% of adolescent female daily smokers reporting feeling calmer while smoking compared with 38% of non-daily smokers.[1] An earlier study[2] of over 5,000 adolescent boys found that 68% of those who smoked agreed that smoking helped them feel more at ease, and 79% agreed that smoking can help people when they feel nervous or embarrassed. There was also some evidence suggesting that these percentages increase with age.

Studies in adults have found similar results. In one study of 600 adult smokers,[3] 53% agreed strongly and 38% agreed mildly with the statement 'smoking can help people relax', whilst only 2% disagreed strongly. Another typical study[4] asked a representative sample of over 500 British smokers how likely they thought that they would be able to do various things (eg relax, concentrate or feel confident with other people) if they:

- carried on smoking
- cut down their smoking by half, or
- stopped smoking altogether.

The response most strongly determined by smoking status was the ability to relax: 77% felt they would be able to relax if they continued smoking and only 8% said they would not, whereas only 40% felt they would be able to relax if they quit, and 38% felt they would be unlikely to be able to relax if they stopped smoking. Other factors were perceived to be less likely to be affected by smoking status: for example, 65% felt they would be able to concentrate if they continued smoking, whilst 56% felt they would be able to concentrate if they quit. Even in studies which failed to find a 'sedative' factor in the questionnaire structure, the questionnaire items which appeared to relate most strongly to negative affect reduction received very high agreement ratings. For example, in the study by Russell *et al*,[5] 73% of a sample of smokers and 91% of a smokers' clinic sample stated that they smoked more when they were worried. This can be compared with only 34% and 57% of the same two samples who felt that smoking helps to keep them going when they are tired, and only 13% and 11% who felt more attractive to the opposite sex when smoking.

Although these studies have identified several reasons given by smokers as to why they smoke, they have not been designed to compare the relative importance of different smoking motives. However, the effect for which smokers most consistently say they smoke is the alleviation of an unpleasant mood state in which they feel tense, irritable and miserable. Adult smokers report this reason for smoking more frequently than saying that smoking helps them to think and concentrate.[5] Interestingly, however, many of the studies described above found that non-smokers also perceived smoking as having a mood regulating effect. Indeed, studies of beliefs about smoking in children have found that children who have never smoked already perceived smoking as having beneficial effects on mood. For example, in the 1994 survey by the Office of Population, Censuses and Surveys of almost 3,000 11–15 year olds in England,[6] 58% of never-smokers agreed with the statement, 'smoking helps people relax if they feel nervous'. Of those who were already weekly smokers, 89% agreed with this statement. However, only 25% of never-smokers and 32% of regular smokers agreed with the statement that 'smokers stay slimmer than non-smokers'. This suggests that some beliefs about smoking may be learned by methods other than personal experience, and raises the question of whether smoking or nicotine really does produce a positive effect on mood and/or cognitive performance.

Laboratory studies of the effect of nicotine and smoking on mood

Since smokers frequently state that they feel more inclined to smoke when they feel tense, embarrassed, depressed, angry or bored, and readily agree with statements to the effect that smoking can help reduce these unpleasant emotions, it would be expected to be fairly straightforward to design and conduct a laboratory study showing that smoking a cigarette (or receiving nicotine some other way) improves the smoker's mood relative to some control procedure.

However, numerous laboratory studies have failed to detect any mood enhancing effects of smoking or nicotine. In one fairly typical study,[7] 16 overnight-deprived smokers smoked five medium-nicotine cigarettes or five nicotine-free cigarettes, each separated by 30 minutes. The subjects' mood was assessed by a Feeling State Questionnaire (0–10 ratings of strength of 19 subjective states including tension, happiness, relaxation and worry), completed before the first and last cigarettes and after the second and last cigarettes. The only one of these 19 mood items which showed a significant improvement on nicotine cigarettes compared with nicotine-free cigarettes was drowsiness which was reduced to a greater degree by the nicotine cigarettes. This study found that the first cigarette of the day produced a mild stimulant effect, and that this effect was maintained by subsequent cigarettes. However, no effects on dysphoric mood were detected.

A review of the area of smoking and affect regulation[8] concluded that 'nicotine reduces anxiety and negative affect in chronic smokers'. Eight studies were cited which failed to find beneficial effects of nicotine on mood, but it was suggested that in these studies either the smokers of high-nicotine cigarettes received too much nicotine, or that the mood enhancing effects occur only when the smoker is exposed to mild to moderate anticipatory anxiety. This might explain why studies with no stressor[7] and studies involving brief periods of high stress[9] do not generally find a calming effect of smoking.

On the other hand, the few laboratory studies which found improvements in mood following smoking nicotine cigarettes have generally studied nicotine-deprived smokers. It therefore remains possible that any effects are attributable to withdrawal relief rather than to mood enhancement *per se*.[10] Similarly, many of the studies requiring participants to smoke a cigarette have not included adequate control conditions, with participants also smoking a denicotinised cigarette in a double-blind manner. Any effects detected with the weaker designs could easily be attributable to the expectations or demand characteristics of the experimental situation.

Some studies have administered controlled doses of nicotine in a placebo-controlled, double-blind fashion to smokers and non-smokers using routes other than inhalation. One study[11] found that subcutaneous injections of nicotine produced no mood enhancement in either 24-hour deprived smokers or never-smokers, with the mood of the latter worsening on the higher nicotine dose (0.6 mg). These results are consistent with those of others[12,13] who have found that nicotine administered to non-smokers via a nasal spray produced dose-dependent increases in ratings of feeling 'jittery', with no beneficial effects on mood. Similarly, it has been found that intravenous nicotine increased anxiety relative to placebo in non-smoking patients with Alzheimer's disease, and that a moderate dose increased ratings of tension, depression and confusion over those reported with a lower dose (no placebo was given in this study) in healthy non-smoking volunteers.[14] It is possible that the worsening of mood at high nicotine doses in these studies is due to mild overdosing, but the complete absence of any beneficial effects at lower doses, either before or after a stressor,[11] shows that there is no primary mood enhancing effect of nicotine. It seems most likely that the belief that smoking improves mood develops from the repeated experience of mood worsening during periods of abstinence (via nicotine withdrawal), rather than from a consistent effect of smoking improving mood above baseline (never-smoker) levels.

✦

Laboratory studies of nicotine and smoking effects on cognitive performance

The effects of nicotine and smoking on human performance were comprehensively reviewed by Heishman *et al.*[15] They concluded that in non-abstinent smokers and non-smokers, only finger tapping speed and similar motor responses were reliably enhanced by nicotine, and that there were no reliable improvements on tasks with a greater cognitive loading. On the other hand, nicotine was found to have more widespread performance enhancing effects in abstinent smokers. This was interpreted as evidence that nicotine can reverse withdrawal-induced deficits in several areas of performance.

Since that review, however, a number of studies have provided further evidence suggestive of absolute improvements in cognitive performance in humans following nicotine absorption. Low doses of subcutaneous nicotine in non-smokers were found to produce faster reaction times in attentional tasks.[16,17] Similarly, low doses of nicotine improved recognition memory in non-smokers,[13] and smoking a cigarette produced similar improvements in rapid information processing whether the smokers were deprived of nicotine for one or 12 hours.[18] Heishman updated his earlier review in the light of some of these more recent studies and concluded that this new evidence indicates that nicotine produces true enhancement of certain aspects of attention and cognition.[19] Further evidence of improvements in aviation simulator flight performance[20] and attention tasks[21] in non-smokers following placebo-controlled nicotine administration lends further support to this conclusion, which is also consistent with the evidence from studies in rats.[22,23] However, the magnitude of this effect of nicotine is small, comparable to the effects obtained by consuming caffeinated beverages.[24-26]

Other evidence on the psychological effects of nicotine and its withdrawal

Consistent with the laboratory studies, there is good evidence that when smokers attempt to abstain for short periods outside the laboratory environment, their mood and cognitive performance temporarily worsens.[27,28] That this nicotine withdrawal syndrome can produce significant mental impairment is supported by a recent study[29] which found consistently higher numbers of non-fatal accidents at work reported in the UK on national No Smoking Day than on normal working days over a 10-year period.

There is also clear evidence that smokers tend to be less mentally healthy than never- or ex-smokers, rather than being happier or calmer people.[30] Indeed, within countries such as the UK and the US, tobacco use is much more prevalent amongst people with serious mental disorders such as schizophrenia[30] or among those incarcerated in prisons.[31] This in itself should not be interpreted as strong evidence that smoking causes stress or poor mental health, since it is possible that these individuals use tobacco primarily in an attempt to reduce their stress.[32] However, other evidence is consistent with the idea that being a smoker may increase stress and that, in the longer term, quitting actually decreases stress.

When the mood of smokers is measured before and after quitting, their mood typically worsens during the first few days of abstinence, and then returns to previous levels within about four weeks. In studies which have continued to measure these quitters' mood over a longer period, the general finding is that it continues to improve above the level when they were smoking.[33] Similarly, it has been found that smokers who manage to quit for six months report a steady reduction in stress from the first month of abstinence, such that six months after quitting their stress levels are lower than when they smoked.[34]

Additional evidence that smoking may actually *increase* stress comes from studies of the daily pattern of stress change in smokers. Parrott *et al* have conducted a number of studies[35,36] which demonstrated that smokers' stress levels during a normal smoking day increase rapidly between cigarettes, the net effect being large fluctuations in mood and greater total stress than occurs in non-smokers.

Conclusions

It is evident that smokers perceive that smoking helps alleviate negative mood states, but the available evidence suggests that the only negative mood state so alleviated is that resulting directly from the nicotine dependence itself. Thus, the nicotine in tobacco relieves nicotine withdrawal symptoms, but does not have mood enhancing properties in non-addicted individuals. If anything, the experience of being addicted to tobacco appears to add to, rather than relieve, stress in the everyday lives of smokers.

Paradoxically, although relatively few smokers report that they smoke primarily to help them think and concentrate, the evidence suggests that nicotine can improve certain aspects of cognitive performance, even in non-addicted individuals. The magnitude of this effect is however small.

3.2 The nicotine withdrawal syndrome

A drug withdrawal syndrome is a collection of signs and symptoms caused by abstinence from use of a drug to which there has been physiological adaptation. The symptoms should be temporary because, after a period of sustained abstinence, the body should revert to a normal, drug-free state.[37]

Abstinence from cigarette smoking is associated with a characteristic set of signs and symptoms which may be labelled the 'cigarette withdrawal syndrome'. Determining whether this syndrome is due specifically to nicotine, as opposed to the loss of other aspects of smoking, requires a demonstration that it is prevented by ingestion of nicotine from another source (eg nicotine chewing gum) or that the syndrome occurs as a result of abstinence from other forms of nicotine intake. Such a demonstration is undermined by the fact that alternative nicotine delivery systems generally result in lower levels of nicotine intake than smoking and also provide nicotine much more slowly. Therefore, even when a particular element of the cigarette withdrawal syndrome is not reliably alleviated by, for example, nicotine gum, it may still represent a nicotine withdrawal state.[37] However, there is strong evidence that nicotine replacement reduces the severity of the cigarette withdrawal syndrome in general, and specific elements in particular.[38–40]

The major changes in mood, physical symptoms and physiological variables that have been reliably shown to follow abstinence from smoking are listed in Table 3.1. The most prominent are symptoms of anxiety, restlessness, poor concentration and irritability or aggression,

Table 3.1. Major signs and symptoms of cigarette withdrawal.

Symptom	Duration[27]	Reduced by NRT	Predicts relapse[41]	Incidence Self-quitters[27] (%)	Clinic patients[42] (%)
Irritability/aggression	<4 weeks	Yes	No	38	80
Depression	<4 weeks	Yes	Yes	31	60[43]
Anxiety	<2 weeks	Yes	No	49	87
Restlessness	<2 weeks	Yes	No	46	71
Poor concentration	<1 week	Yes	No	43	73
Increased appetite	>10 weeks	Yes	No	53	67
Urges to smoke	>2 weeks	Yes	Yes	37	62
Night-time awakenings	<1 week	nk	No	39	24
Decreased heart rate	>10 weeks	Yes	nk	61	79
Decreased adrenaline	<2 weeks[39]	nk	nk	nk	nk
Decreased cortisol	nk	nk	nk	nk	nk

nk = not known; NRT = nicotine replacement therapy.

their relative frequency varying between different smoking populations. The duration of these responses after withdrawal, measured in terms of the time for which they have been shown to be significantly different from pre-abstinence levels, is predominantly four weeks or less, though some effects, such as the increase in appetite, are more protracted. Table 3.1 also identifies those characteristics for which there is consistent evidence of alleviation by nicotine replacement. Studies carried out into how far these responses cohere into a single syndrome suggest that, whilst mood changes tend to go together, increased appetite does not correlate well with other symptoms.[44] Tate *et al*[45] have also reported that the presentation of withdrawal symptoms within smokers appears to be relatively consistent over separate, closely spaced abstinence periods, with the possible exception of increased appetite.

There has been some debate as to whether cravings or urges to smoke are also part of this withdrawal syndrome. A powerful case can be made for their inclusion since there is now good evidence that these cravings are reduced by nicotine replacement[38] and are correlated with other elements of the withdrawal syndrome.[46] In fact, there is also evidence that urges to smoke are probably the single most important element of the withdrawal syndrome, in that they are most clearly predictive of subsequent relapse to smoking.[47–49]

There has also been debate about whether increased anxiety should be included as a withdrawal symptom. Recent research[50] suggests that it should not because it has been found that anxiety levels actually fall rather than rise among totally abstaining smokers (as opposed to those who might have had minor lapses). Some studies have reported an initial elevation in anxiety after stopping smoking, but this is short-lived and followed by a drop to below the levels while smoking.[41] It has been argued that the increase in anxiety observed in some studies on cessation of smoking is a psychological response to the attempt to stop, which is made worse when that attempt is not being wholly successful.[50] This is an issue that requires clarification.

The overall conclusion that withdrawal from cigarettes results in symptoms and signs that are reversed by nicotine strongly implicates nicotine dependence as at least one major component of dependence on smoking. The concept of the withdrawal symptoms also has impor-tant implications for our understanding of the psychological basis for continued smoking.

3.3 Psychological dependence on nicotine and smoking

The development of dependence is an important component of the psychological effects of any drug, and there is abundant evidence that nicotine dependence develops in cigarette smokers. Some of this is derived from observational studies of the trends and patterns of nico-tine consumption by smokers, whilst other evidence comes from studies of measures of dependence within smokers. In some of these studies there is an explicit theoretical rationale for the link between the measurement used and the core concept of compulsive use, whereas in others the link is based on unstated assumptions. The evidence is illustrated as follows.

Consumption of nicotine by smokers

Average consumption of cigarettes by male and female smokers in Britain is currently 16 and 14 cigarettes per day respectively, a figure which has changed little over the past 20 years.[51] The amount of nicotine ingested from each cigarette varies considerably between individuals, but has been estimated at approximately 1.0–1.5 mg.[52,53] These estimates, and also estimates derived from measurements of plasma cotinine concentrations,[54] suggest a daily intake of nicotine in smokers of approximately 16–24 mg per day in men and 14–21 mg per day in women. In the absence of constraints by restrictions such as workplace smoking bans, most smokers smoke regularly throughout the day and 34% have their first cigarette within 15 minutes of waking.[51] Daily cigarette consumption has been shown in several studies to be consistent over periods of months.[55–57] The fact that smoking patterns are so consistent and stable, both in the general population of smokers over time and within individual smokers, implies some degree of dependence on cigarettes.

Markers and measures of dependence in smokers

Daily cigarette consumption. A frequently used marker of dependence is daily cigarette consumption, on the assumption that the more cigarettes smoked per day the harder people should find it to stop. This might reflect several mechanisms, in that a high level of nicotine intake could:

- lead to more pronounced neuro-adaptation
- reflect a greater constitutional need for nicotine, or
- reflect a more highly learned and deeply ingrained habit.

Whatever the mechanism, there is clear evidence that those who smoke more cigarettes per day are less likely to be able to stop.[55,58,59]

Biochemical markers of nicotine intake. Biochemical markers have also been used as indices of dependence. The most widely used is the concentration of the nicotine metabolite, cotinine, in saliva. This has the advantages, first, of being easily obtained and reflecting nicotine intake over a period of days because of its relatively long half-life,[57] and secondly, that cotinine more accurately reflects nicotine intake than daily cigarette consumption. Above about 10 cigarettes per day, the correlation between daily cigarette consumption and nicotine intake is weak,[57] probably because of the important influence of puffing and inhalation patterns. In fact, it is possible in principle to obtain as much nicotine from five cigarettes per day as most smokers achieve with 30 per day (see Chapter 6). The level of cotinine in saliva has been shown to predict success of attempts to stop smoking.[60]

Questionnaire methods of measuring dependence on nicotine. Questionnaire methods have also been used extensively to measure nicotine dependence. Probably the most widely used is the Fagerström Test for Nicotine Dependence (FTND) (see Table 3.2),[61] a shortened version of the Fagerström Tolerance Questionnaire (FTQ).[62] The measure combines

Table 3.2. The Fagerström test for nicotine dependence.

Question	Answer	Score
How soon after you wake up do you smoke your first cigarette	Within 5 minutes	3
	6–30 minutes	2
	31–60 minutes	1
	>60 minutes	0
Do you find it difficult to refrain from smoking in places where it is forbidden?	Yes	1
	No	0
Which cigarette would you hate to give up most?	The first one in the morning	1
	Others	0
How many cigarettes per day do you smoke?	≤10	0
	11–20	1
	21–30	2
	≥31	3
Do you smoke more frequently during the first hours after waking than during the rest of the day?	Yes	1
	No	0
Do you smoke if you are so ill that you are in bed most of the day?	Yes	1
	No	0

Note: scores are totalled to yield a single value.

an index of consumption (cigarettes per day) with difficulty tolerating reduced nicotine levels:

- time to first cigarette of the day
- smoking even when ill
- smoking more in the morning
- difficulty not smoking in no smoking areas
- which cigarette would hate to give up.

A two-item test, the Heaviness of Smoking Index (HSI), which measures the number of cigarettes per day and the time to the first cigarette of the day, has also been examined.[63] The extent to which the FTND measures a single construct has been called into question, as has the theoretical underpinning of the questionnaire.[55,64] A threshold of 7 on the FTQ and 6 on the FTND is generally used to divide smokers into high and low dependence categories. Because the FTQ, FTND and HSI have been shown in many studies to predict failure of attempts to stop smoking,[65,66] they are held to provide evidence of dependence. However, in terms of quantifying this, it is not clear how far these measures are any improvement over simply measuring the number of cigarettes smoked per day.[55]

Another questionnaire measure of dependence is the dependence subscale of the Smoking Motivation Questionnaire[41] which focuses on subjective feelings of dependence and cravings during abstinence. This scale has been shown to predict the severity of urges to smoke during quit attempts.[41] Subjective dependence, measured in terms of perceived difficulty maintaining abstinence, has also been shown to predict failure of quit attempts.[65] In Britain, 32% of smokers report that they would find it 'very difficult' to go without smoking for a whole day.[51]

Other measures of dependence have been derived from the American Psychiatric Association's diagnostic criteria for substance dependence, the Diagnostic and Statistical Manual of Mental Disorders (DSM)-III,[67] and DSM-IV,[68] and from the World Health Organization's International Classification of Diseases (ICD)-10 criteria[69] (see Chapter 4 for details):

- Hughes et al[70] reported that 90% of a general population sample of 1,006 middle-aged male smokers fulfilled DSM-III criteria for dependence, while 36% had an FTQ score of 7 or more. However, concordance between the two measures was low.

- Nelson and Wittchen[71] used the Composite International Diagnostic Interview (CIDI) (a questionnaire based on DSM-IV) to assess dependence in a sample of 3,021 smokers in Germany aged 14–24 years. They defined 52% of smokers as dependent, most frequently on the grounds of failure of attempts to control or cease use.

- Kandel et al[72] used items derived from DSM-IV criteria to study drug dependence in a sample of 87,915 adults, and concluded that nicotine was more highly addictive than alcohol, cocaine or marijuana.

- Kawakami et al[73] examined lifetime prevalence of dependence in male ever-smokers aged 35 or over in Japan using the CIDI to measure dependence according to ICD-10, DSM-IIIR and DSM-IV criteria. Estimates of lifetime prevalence of dependence by these criteria were 42%, 26% and 32%, respectively. They also measured dependence using the FTQ, and found that 19% had a FTQ score of 7 or more. The ICD-10 diagnosis was significantly associated with failure to stop smoking.

- Kawakami et al[74] have recently developed a new 10-item questionnaire, the Tobacco Dependence Screener, for assessing tobacco dependence according to ICD-10, DSM-IIIR and DSM-IV criteria. This instrument has been shown to have a higher coherence (intercorrelations among items) than the FTQ, to be associated with cigarette consumption, and to predict failure to stop smoking.

Conclusions

There is considerable evidence from questionnaire studies that certain smoking characteristics can predict failure in an attempt to give up smoking, and thus to identify in broad terms dependence or addiction to smoking. Markers of nicotine consumption indicate that smokers maintain a relatively consistent nicotine intake, and that failure to maintain that intake results in symptoms of nicotine withdrawal. Studies of mood effects indicate that the major psychological motivation to smoke is the avoidance of negative mood states caused by withdrawal of nicotine. It is therefore evident that nicotine plays a fundamental role in smoking behaviour.

References

1 McNeill AD, Jarvis M, West, R. Subjective effects of cigarette smoking in adolescents. *Psychopharmacology* 1997; **92**: 115–7.

2 Brynner JM. *The young smoker*. London: HMSO, 1969.

3 McKennell AC, Thomas RK. *Adults and adolescents smoking habits and attitudes*. London: British Ministry of Health, 1967.

4 Marsh A, Mathieson J. *Smoking attitudes and behaviour*. London: HMSO, 1983.

5 Russell MAH, Peto J, Patel UA. The classification of smoking by factorial structure of motives. *J R Stat Soc* 1974; **137**: 313–33.

6 Diamond A, Goddard E. *Smoking among secondary school children in 1994*. London: HMSO, 1994.

7 Meliska CJ, Gilbert DG. Hormonal and subjective effects of smoking the first five cigarettes of the day: a comparison in males and females. *Pharmacol Biochem Behav* 1991; **40**: 229–35.

8 Gilbert DG, Wesler R. Emotion, anxiety and smoking. In: Ney T, Gale A (eds). *Smoking and human behavior*. Chichester: Wiley, 1989.

9 Fleming SE, Lombardo TW. Effects of cigarette smoking on phobic anxiety. *Addict Behav* 1987; **12**: 195–8.

10 Pomerleau CS, Pomerleau OF. Cortisol response to a psychological stressor and/or nicotine. *Pharmacol Biochem Behav* 1990; **35**: 211–3.

11 Foulds J, Stapleton JA, Bell N, Swettenham J, *et al.* Mood and physiological effects of subcutaneous nicotine in smokers and never-smokers. *Drug Alcohol Depend* 1997; **44**: 105–15.

12 Perkins KA, Grobe JE, Epstein LH, Caggiula A, *et al.* Chronic and acute tolerance to subjective effects of nicotine. *Pharmacol Biochem Behav* 1993; **45**: 375–81.

13 Perkins KA, Grobe JE, Fonte C, Goettler J, *et al.* Chronic and acute tolerance to subjective, behavioral and cardiovascular effects of nicotine in humans. *J Pharmacol Exp Ther* 1994; **270**: 628–38.

14 Newhouse PA, Sunderland T, Narang PK, Mellow AM, *et al.* Neuroendocrine, physiologic, and behavioral responses following intravenous nicotine in non-smoking healthy volunteers and in patients with Alzheimer's disease. *Psychoneuroendocrinology* 1990; **15**: 471–84.

15 Heishman SJ, Taylor RC, Henningfield JE. Nicotine and smoking: a review of effects on human performance. *Exp Clin Psychopharmacol* 1994; **2**: 345–95.

16 Le Houezec J, Halliday R, Benowitz NL, Callaway E, *et al.* Low dose of subcutaneous nicotine improves information processing in non-smokers. *Psychopharmacology* 1994; **114**: 628–34.

17 Foulds J, Stapleton J, Swettenham J, Bell N, *et al.* Cognitive performance effects of subcutaneous nicotine in smokers and never-smokers. *Psychopharmacology* 1996; **127**: 31–8.

18 Warburton DM, Arnall C. Improvements in performance without nicotine withdrawal. *Psychopharmacology* 1994; **115**: 539–42.

19 Heishman SJ. What aspects of human performance are truly enhanced by nicotine? *Addiction* 1998; **93**: 317–20.

20 Mumenthaler DM, Taylor JL, O'Hara R, Yesevage JA. Influence of nicotine on simulator flight performance in non-smokers. *Psychopharmacology* 1998; **140**: 38–41.

21 Levin ED, Conners CK, Silva D, Hinton SC, *et al.* Transdermal nicotine effects on attention. *Psychopharmacology* 1998; **140**: 134–41.

22 Mirza NR, Stolerman IP. Nicotine enhances sustained attention in the rat under specific task conditions. *Psychopharmacology* 1998; **138**: 266-74.

23 Levin ED, Simon BB. Nicotine acetylcholine involvement in cognitive function in animals. *Psychopharmacology* 1998; **138**: 217–30.

24 Kerr JS, Sherwood N, Hindmarch I. Separate and combined effects of the social drugs on psychomotor performance. *Psychopharmacology* 1991; **104**: 113–9.

25 Jarvis MJ. Does caffeine intake enhance absolute levels of cognitive performance? *Psychopharmacology* 1993; **110**: 45–52.

26 Cohen C, Pickworth WB, Bunker EB, Henningfield JE. Caffeine antagonises EEG effects of tobacco withdrawal. *Pharmacol Biochem Behav* 1994; **47**: 919–26.

27 Hughes JR. Tobacco withdrawal in self-quitters. *J Consult Clin Psychol* 1992; **60**: 689–97.

28 Shiffman S, Paty JA, Gnys M, Kassel JD, Elash C. Nicotine withdrawal in chippers and regular smokers: subjective and cognitive effects. *Health Psychol* 1995; **14**: 301–9.

29 Waters AJ, Jarvis MJ, Sutton SR. Nicotine withdrawal and accident rates. *Nature* 1998; **394**: 137.

30 Foulds J. The relationship between tobacco use and mental disorders. *Curr Opin Psychiatry* 1999; **12**: 303–6.

31 Colsher PL, Wallace RB, Loeffelholz PL, Sales M. Health status of older male prisoners: a comprehensive survey. *Am J Pub Health* 1992; **82**: 881–4.

32 Foulds J. Detrimental effects of nicotine on mood? *Addiction* 1994; **89**: 136–7.

33 Hughes JR, Higgins ST, Hatsukami D. Effects of abstinence from tobacco. In: Kowslowski LT, Annis HM (eds). *Recent advances in alcohol and drug problems.* New York: Plenum, 1990.

34 Cohen S, Lichtenstein E. Perceived stress, quitting smoking, and smoking relapse. *Health Psychol* 1990; **9**: 466–78.

35 Parrott AC. Stress modulation over the day in cigarette smokers. *Addiction* 1995; **90**: 233–44.

36 Parrott AC. Nesbitt's Paradox resolved? Stress and arousal modulation during cigarette smoking. *Addiction* 1998; **93**: 27–39.

37 West R, Gossop M. A comparison of withdrawal symptoms from different drug classes. *Addiction* 1994; **89**: 1483–9.

38 Russell MAH, Stapleton JA, Feyerabend C, Wiseman SM, *et al.* Targeting heavy smokers in general practice: randomised controlled trial of transdermal nicotine patches. *Br Med J* 1993; **306**: 1308–12.

39   Gross J, Stitzer ML. Nicotine replacement: ten-week effects on tobacco withdrawal symptoms. *Psychopharmacology* 1989; **98**: 334–41.

40   Hughes JR, Gust SW, Skoog K, Keenan RM, Fenwick JW. Symptoms of tobacco withdrawal. A replication and extension. *Arch Gen Psychiatry* 1991; **48**: 52–9.

41   Hughes JR, Higgins ST, Bickel WK. Nicotine withdrawal versus other drug withdrawal syndromes: similarities and dissimilarities. *Addiction* 1994; **89**: 1461–70.

42   Hughes JR, Hatsukami DK. Signs and symptoms of tobacco withdrawal. *Arch Gen Psychiatry* 1986; **43**: 289–94.

43   West R, Russell M, Jarvis M, Pizzie T, Kadam B. Urinary adrenaline concentrations during 10 days of smoking abstinence. *Psychopharmacology* 1984; **84**: 141–2.

44   West R, Russell M. Pre-abstinence smoke intake and smoking motivation as predictors of severity of cigarette withdrawal symptoms. *Psychopharmacology* 1985; **87**: 334–6.

45   Tate JC, Pomerleau OF, Pomerleau CS. Temporal stability and within-subject consistency of nicotine withdrawal symptoms. *J Subst Abuse* 1993; **5**: 355–63.

46   Zinser MC, Baker TB, Sherman JE, Cannon DS. Relation between self-reported affect and drug urges and cravings in continuing and withdrawing smokers. *J Abn Psychol* 1992; **101**: 617–29.

47   Killen JD, Fortmann SP, Newman B, Varady A. Prospective study of factors influencing the development of craving associated with smoking cessation. *Psychopharmacology* 1991; **105**: 191–6.

48   Swan GE, Ward MM, Jack LM. Abstinence effects as predictors of 28-day relapse in smokers. *Addict Behav* 1996; **21**: 481–90.

49   West R, Hajek P, Belcher M. Severity of withdrawal symptoms as a predictor of outcome of an attempt to quit smoking. *Psychol Med* 1989; **19**: 981–5.

50   West R, Hajek P. What happens to anxiety levels on giving up smoking? *Am J Psychiatry* 1997; **154**: 1589–92.

51   Office for National Statistics. *Living in Britain: results from the 1996 General Household Survey.* London: The Stationary Office, 1997.

52   Benowitz NL, Jacob P. Daily intake of nicotine during cigarette smoking. *Clin Pharmacol Ther* 1984; **35**: 499–504.

53   Perez-Stable EJ, Herrera B, Jacob P, Benowitz NL. Nicotine metabolism and intake in black and white smokers. *JAMA* 1998; **280**: 152–6.

54   Benowitz NL, Jacob P. Metabolism of nicotine to cotinine studied by a dual stable isotope method. *Clin Pharmacol Ther* 1994; **56**: 483–93.

55   Etter JF, Duc TV, Perneger TV. Validity of the Fagerstrom test for nicotine dependence and of the Heaviness of Smoking Index among relatively light smokers. Review. *Addiction* 1999; **94**: 269–81.

56   West R, McEwen A, Bolling K. *Smoking cessation and harm minimisation strategies in the general population.* London: Health Education Authority, 1999.

57   Kemmeren JM, van Poppel G, Verhoef P, Jarvis MJ. Plasma cotinine: stability in smokers and validation of self-reported smoke exposure in nonsmokers. *Environ Res* 1994; **66**: 235–43.

58   Hymowitz N, Cummings M, Hyland A, Lynn WR, *et al.* Predictors of smoking cessation in a cohort of adult smokers followed for five years. *Tob Control* 1997; **6**: S57–62.

59   Senore C, Battista RN, Shapiro SH, Segnan N, *et al.* Predictors of smoking cessation following physicians' counselling. *Prev Med* 1998; **27**: 412–21.

60   West R. Nicotine is addictive: the issue of free choice. In: Clarke PBS, Quik M, Adlkofer F, Thurau K (eds). *Effects of nicotine on biological systems II.* Basel: Birkhauser, 1995: 265–72.

61 Heatherton TF, Kozlowski LT, Frecker RC, Fagerström KO. The Fagerström Test for Nicotine Dependence: a revision of the Fagerström Tolerance Questionnaire. *Br J Addict* 1991; **86**: 1119–27.

62 Fagerström KO, Schneider NG. Measuring nicotine dependence: a review of the Fagerström Tolerance Questionnaire. *J Behav Med* 1989; **12**: 159–82.

63 Heatherton TF, Kozlowski LT, Frecker RC, Rickert WS, Robinson J. Measuring the heaviness of smoking: using self-reported time to the first cigarette of the day and number of cigarettes smoked per day. *Br J Addict* 1989; **84**: 791–9.

64 Haddock CK, Lando H, Klesges RC, Talcott W, Renaud EA. A study of the psychometric and predictive properties of the Fagerstrom Test for Nicotine Dependence in a population of young smokers. *Nicotine Tob Res* 1999; **1**: 59–66.

65 Gritz ER, Carr CR, Marcus AC. Unaided smoking cessation: Great American Smokeout and New Year's Day quitters. *J Psychosocial Oncol* 1988; **6**: 217–34.

66 Kozlowski LT, Porter CQ, Orleans T, Pope MA, Heatherton T. Predicting smoking cessation with self-reported measures of nicotine dependence: FTQ, FTND and HSI. *Drug Alcohol Depend* 1994; **34**: 211–6.

67 American Psychiatric Association. *Diagnostic and Statistical Manual of Mental Disorders*, 3rd edn (revised). Washington: APA, 1987.

68 American Psychiatric Association. *Diagnostic and Statistical Manual of Mental Disorders*, 4th edn. Washington: APA, 1995.

69 World Health Organization. *International Classification of Diseases and related Health Problems*, 10th revision. Geneva: WHO, 1992.

70 Hughes JR, Gust SW, Pechacek TF. Prevalence of tobacco dependence and withdrawal. *Am J Psychiatry* 1987; **144**: 205–8.

71 Nelson CB, Wittchen HU. Smoking and nicotine dependence. Results from a sample of 14- to 24-year-olds in Germany. *Eur Addict Res* 1998; **4**: 42–9.

72 Kandel D, Chen K, Warner LA, Kessler RC, Grant B. Prevalence and demographic correlates of symptoms of last year dependence on alcohol, nicotine, marijuana and cocaine in the US population. *Drug Alcohol Depend* 1997; **44**: 11–29.

73 Kawakami N, Takatsuka N, Shimizu H, Takai A. Life-time prevalence and risk factors of tobacco/nicotine dependence in male ever-smokers in Japan. *Addiction* 1998; **93**: 1023–32.

74 Kawakami N, Takatsuka N, Inaba S, Shimizu H. Development of a screening questionnaire for tobacco/nicotine dependence according to ICD-10, DSM-IIIR and DSM-IV. *Addict Behav* 1999; **24**: 155–66.

4 | Is nicotine a drug of addiction?

4.1 The definition of addiction and dependence
4.2 Does nicotine use through smoking meet standard diagnostic criteria for addiction?
4.3 The history of social, cultural and political responses to nicotine addiction in Britain
4.4 How does nicotine addiction compare with addiction to other drugs?
4.5 Relevance to society of recognition of nicotine as an addictive drug

4.1 The definition of addiction and dependence

'Addiction' and 'dependence' are terms whose definition has a social as well as a scientific dimension. In principle, they may be distinguished, but in practice such a distinction serves little purpose and the terms are used interchangeably here. They are socially and scientifically defined in that their meaning can be, and has been, changed to reflect changing perceptions rather than to identify unequivocally an invariant, objectively definable entity. Under the current definition, the terms refer to a situation in which a drug or stimulus has unreasonably come to control behaviour.[1,2]

This definition is very different from that used in the past and to which the general public mostly subscribe.[3] This earlier and popular view is that addiction refers to a state in which an individual needs to continue to take a drug in order to stave off unpleasant or dangerous withdrawal effects. The main shortcoming of this approach to defining addiction is that it addresses just one aspect of a wider problem. Certainly, many drug addicts experience withdrawal discomfort when they abstain, and this provides an important motive for continuing to use the drug. However, it has also long been recognised that this motive plays a relatively modest role in the apparently unreasonable continued use of a drug, despite protestations of users that they want to stop, and despite the harm their drug use is doing both to them and to those around them. Individuals given morphine for pain relief may experience withdrawal symptoms when it is withdrawn but do not become compulsive users, yet individuals attempting to stop using drugs, including nicotine, continue to relapse at a high rate long after withdrawal symptoms have resolved. Moreover, controlling withdrawal symptoms alone is not necessarily sufficient to prevent relapse to drug use.[4]

Another outmoded feature of the definition of addiction is inclusion of the concept of intoxication.[3] Under this view, addictive drugs lead to changes in users' psychological state, leaving some degree of impairment. This feature no longer appears in any official definition because it is apparent that it is neither a necessary nor a sufficient condition for compulsive, harmful drug seeking.[5] Many cannabis and alcohol users become intoxicated but do not develop dependence, while cocaine and amphetamine use can be compulsive without performance being noticeably impaired. Definitions of dependence or addiction to drugs in general have had to develop and evolve over time to take account of changes in these and other relevant concepts.

Three of the most widely used generic criteria for substance dependence are summarised in Table 4.1. These comprise the American Psychiatric Association (APA) Diagnostic and Statistical Manual of Mental Disorders (DSM)-IIIR criteria,[6] the DSM-IV criteria[7] (which superseded DSM-IIIR in 1995), and the World Health Organization International Classification of Diseases (ICD)-10 criteria.[2] The APA criteria are generally much more detailed than ICD-10, but share common concepts of difficulty in controlling the use of the drug, of giving priority to drug use over other important obligations, to continued use of the drug in the knowledge of harmful consequences, and tolerance to the effects of the drug. The criteria are designed to apply generically to substance abuse, but provide a suitable framework for determining the addictive or dependent nature of nicotine and smoking.

4.2 Does nicotine use through smoking meet standard diagnostic criteria for addiction?

This question is addressed by assessing nicotine intake through smoking in relation to the specific criteria listed in the DSM-IV or ICD-10 definitions, as follows:

A strong desire to take the drug. This characteristic is included in ICD-10 but not in DSM-IV. In addition, DSM-IV does not include craving or a similar item in its list of withdrawal symptoms. As discussed in Chapter 3, the desire to smoke plays a crucial role in relapse of smokers trying to give up smoking, is a manifestation of nicotine withdrawal and is clearly related to underlying dependence on nicotine.

Table 4.1. Summary of Diagnostic and Statistical Manual of Mental Disorders (DSM-IIIR), DSM-IV and International Classification of Diseases (ICD)-10 criteria for substance dependence.

DSM-IIIR	DSM-IV	ICD-10
At least 3 of:	At least 3 of:	A cluster of behavioural, cognitive and physiological phenomena that develop after repeated substance use and that typically include:
Substance often taken in larger amounts or over a longer period than intended	Substance often taken in larger amounts or over a longer period than intended	A strong desire to take the drug
Persistent desire or one or more unsuccessful efforts to cut down or control use	Persistent desire or unsuccessful efforts to cut down or control use	Difficulty controlling use
A great deal of time spent in activities necessary to get the substance, use the substance or recover from its effects	A great deal of time spent in activities necessary to obtain the substance, use the substance or recover from its effects	
Frequent intoxication or withdrawal symptoms when expected to fulfil major role obligations or when substance use is physically hazardous		
Important social, occupational or recreational activities given up or reduced because of substance use	Important social, occupational or recreational activities given up or reduced because of substance use	A higher priority given to drug use than to other activities and obligations
Continued substance use despite knowledge of having a persistent or recurrent social, psychological or physical problem that is caused or exacerbated by the use of the substance	Continued substance use despite knowledge of having a persistent or recurrent social, psychological or physical problem that is caused or exacerbated by the use of the substance	Persisting in use despite harmful consequences
Tolerance: need for markedly increased amounts of the substance to achieve intoxication or desired effect or markedly diminished effect with continued use of the same amount	Tolerance: need for markedly increased amounts of the substance to achieve intoxication or desired effect or markedly diminished effect with continued use of the same amount	Increased tolerance
Characteristic withdrawal symptoms	Withdrawal: the characteristic withdrawal syndrome or the same (or a closely related) substance is taken to relieve or avoid withdrawal symptoms	Sometimes, a physical withdrawal state
Substance often taken to relieve or avoid withdrawal symptoms		

Substance taken in larger amounts or longer than intended. This item is included in DSM-IV, and was originally designed to capture a feature of alcohol dependence.[5] Although the DSM manual states that individuals who smoke are 'likely to find that they use up their supply of cigarettes ... faster than originally intended',[7] this criterion probably does not apply so clearly to nicotine and is not included in ICD-10.

Difficulty in controlling use. The concept of difficulty in controlling smoking is common to both DSM-IV and ICD-10 criteria. Much of the discussion of the psychological effects of nicotine in Chapter 3 relates to this criterion. It is also relevant that surveys of attitudes to smoking find consistently that the majority want to stop smoking (71% in Britain in 1997[8]), and that a similar percentage believe that if they were to try to give up, they would fail.[9] In the UK, about 80% of smokers have made at least one attempt to quit,[9] and some 30% make at least one attempt each year.[10] Only a tiny proportion of quit attempts succeed, so that only approximately 1% of smokers in the UK become long-term ex-smokers each year.[11] Difficulty controlling use of smoking and nicotine is therefore self evident.

A great deal of time is spent in obtaining, using or recovering from effects of substance. This criterion is included in DSM-IV, but not ICD-10. Because cigarettes are legal and relatively inexpensive compared with other drugs of dependence, and because smoking can often be engaged in while doing other things, this criterion is less relevant to smoking.[5] However, the more recent introduction of controls on smoking in the workplace in the UK and many other countries now mean that smoking at work is increasingly an activity which has to be pursued in a designated separate area, whilst the adoption of smoking restrictions in restaurants, bars, public houses and other public areas means that smokers are becoming less able to smoke when and where they choose outside the home. Therefore, as public perceptions and controls on the acceptability of smoking change, smokers are having to spend more time in activities specifically related to smoking; this criterion is gaining relevance as a result.

A higher priority given to drug use than to other activities and obligations. This criterion is included in both DSM-IV and ICD-10. In general, it is not applicable to smoking, because smoking has been a relatively socially acceptable activity in relation to the use of other drugs.[5] However, there are circumstances in which priority is evidently given to smoking over other activities or obligations: for example, when an individual forgoes an activity because it occurs in smoking restricted areas, or when adults smoke at home and expose their children to the risks of passive smoke.

Continued use despite harmful consequences. This criterion is present in DSM-IV and ICD-10. It is clear from surveys that most smokers are aware of the health risks, and that this is one of the main reasons why they wish to stop.[8,10] There is also substantial evidence that the majority of smokers continue to smoke after diagnosis of smoking related disease (see Section 5.2).

Tolerance. This criterion is also present in both DSM-IV and ICD-10. According to the DSM manual,[7] tolerance to nicotine is manifested by absence of nausea, dizziness and other characteristic symptoms rather than by reduction in sought-after effects.

Withdrawal. As discussed in Chapter 3, one of the major motives for continued smoking appears to be the relief of the nicotine withdrawal syndrome.

Discussion

There is a question as to how far the current conceptualisation of dependence as it relates to cigarette smoking reflects one or more than one dimension. Johnson *et al*[12] have reported data suggesting that the DSM-IIIR criteria reflect two factors, one relating to 'general dependence' and the other to 'failed cessation'. Given that the most important marker of dependence is probably the failure of attempts to stop, it may make more sense to consider the dimensions 'subjective' and 'behavioural' dependence, respectively. In relation to this model, behavioural dependence would appear to be the more significant from a public health and medical perspective.

It is also arguable that the social context and specific pharmacology of different substances dictate that different criteria be used as markers of dependence, and therefore that no single set of criteria can provide a universal framework for the definition of dependence or addiction. It is also evident from the above discussion that nicotine and smoking meet both the DSM-IV and ICD-10 criteria for substance dependence. On present evidence, it is reasonable to conclude that nicotine delivered through tobacco smoke should be regarded as an addictive drug, and tobacco use as the means of nicotine self-administration.

4.3 The history of social, cultural and political responses to nicotine addiction in Britain

Different substances have different addiction histories, and the terminology of addiction has also changed over time. The idea of some substances (opium, cocaine, alcohol) as addictive is well over a century old, while the idea of tobacco as an addictive substance is much more recent. This section aims to point to some reasons for this difference. It could be seen in terms of belated understanding or of attempts to hide the truth – but the 'scientific progress' or 'industry conspiracy' arguments mask a greater complexity. We argue here that the significance accorded to conceptual organising tools such as addiction reflects their different social, cultural and policy histories. The point at issue in this section is not whether addiction exists, but rather when, and why, does the concept of nicotine addiction assume significance in both social and policy terms? Here, there are clear differences between drugs or alcohol and smoking, which can be related to a complex of structural and institutional issues and their change over time.

Drugs, alcohol and addiction

For both opium and alcohol, theories of disease were not new even in the 19th century. Rush in the US and Trotter in Britain have been hailed as the 'discoverers' of disease in relation to alcohol addiction, with allied theories relating to opium following later in the century. The disease view of addiction was not 'progress', but rather part of a fundamental paradigm shift which characterised developments in insanity more generally.[13] The language of disease in relation to alcohol was common earlier, in the 18th century. Those 18th century accounts were based on an associationist psychology of habituation, while those of the early 19th century drew on theories of moral insanity as paralysis of the will.[14] What was new in the 19th century was not the theory, but the social and political significance accorded to it.

The temperance movement and teetotalism, with their emphasis on total abstinence, combined with the emergent medical profession to make disease theory the central interpretation of chronic drunkenness. This alliance developed rapidly in the last quarter of the 19th century and also drew on the allied, but distinct anti-opium movement. The primary focus of the latter was on the Indo-Chinese opium trade, but that agitation underlined a distinction between what were termed 'medical' and 'non-medical' uses of opium, the latter being aligned with addiction. In Britain, theories of hereditary degeneration and Lamarckian theories of the inheritance of acquired characteristics were drawn on in the concept of inebriety. This concept was applied to both alcohol and opium, and was organisationally supported by the work of the Society for the Study of Inebriety (founded in 1884).[15] Degenerationist theories were of lessening significance before World War I. Belief in physical causation declined, to be replaced by the concept of 'addiction' to drugs as a 'disease of the will'. Here was the slippage between medical and moral concepts which marked much medical ideology of the time, especially in the sphere of mental illness.

This concept of addiction to drugs came to form the central plank of British drug policy from the 1920s, when the Rolleston Committee report of 1926 defined addiction as a disease requiring medical treatment, including maintenance prescribing. This was a 'harm reduction' approach, considered appropriate for the primarily middle-class drug addict clientele of the inter-war years, and reflecting 'mental medicine's' own shift of emphasis away from the asylum. After World War II, the specific role of psychiatry for both drugs and alcohol and for the treatment of addiction was established through a series of policy documents.[15] Alcohol and drugs were both incorporated within the newer concept of 'dependence' which, with echoes of the 18th century theories, symbolised the enhanced importance of psychology and its interrelationships with psychiatry in the 1970s.[16] For alcohol, these 'disease based' conceptualisations were in place alongside a public health population based approach which emphasised issues of individual responsibility and lifestyle, and was therefore in some senses opposed to the disease model. The concept of addiction to drugs remained more firmly entrenched in the 1980s and 1990s, although changing alliances brought psychiatric and public health/ preventive approaches into closer relationship, in particular through the enhanced focus on treatment as a mode of prevention.

Tobacco and nicotine

Tobacco and nicotine have had a different history. The scientific bodies dealing with inebriety did not so easily encompass the use of tobacco. In 1888, Norman Kerr, President of the Society for the Study of Inebriety, commented:[17]

> Though no defender of tobacco, which it cannot be denied is a mere luxury, injurious to the health of many, even when used in moderation, I am driven to the conclusion that in the philosophical and practical meaning of the term, there is no true tobacco inebriety or mania.

Tobacco remained on the fringes of the late 19th century 'medical model'. Its significance was seen to lie rather within the public health model of the time. The issue of juvenile smoking evoked legislative control through the 1908 Children Act, but here the ideology was that of social hygienist concerns about national efficiency rather than of addiction or inebriety.[18]

Smoking therefore followed a different conceptual and policy route. The National Society of Non-Smokers, the leading anti-smoking organisation in the interwar years, focused not on 'disease' but on the clean air and environmentally harmful aspects of smoking. There was a strong moral emphasis on the selfishness of those who bothered others with their smoke.[19] Medical input was limited, and medical writing on the subject was rare and tended to shy away from placing tobacco in the context of addiction.[20] Unlike alcohol and drugs, tobacco was not poised for the post-World War II 'rediscovery of addiction', not having been conceptualised within the addiction model in the first place.

It was not disease/addiction but another scientific paradigm which came to predominate for smoking and tobacco consumption in the post-war years. This was the role of epidemiology and statistical inference, established in the British context by the work of Doll and Hill. Epidemiology gave smoking a different, albeit still medical, route into policy. Smoking and tobacco became a key issue in the new public health constituency forming itself around epidemiology in the 1950s and 1960s.[21,22] Through the epidemiological route, smoking was set within a model of public health response, stressing taxation, health education and advertising controls – far removed from the disease/dependence basis of psychiatric hegemony for drug and alcohol treatment being established at the same time.

As with opium and alcohol two centuries earlier, concepts of disease and dependence were not absent for smoking and nicotine, but they carried no primary significance, for reasons which will be discussed below. In 1942, Johnston's experiments with nicotine injections to counteract smoking had been carried out on the assumption that smoking tobacco was simply a means of administering nicotine, just as smoking opium was a means of administering morphine.[23] The first Royal College of Physicians reports on smoking in 1962[24] and 1971[25] recognised that smokers might be addicted to nicotine. In line with theories current at that time, the discussion mingled the role of motivation with that of inheritance and personality. In Britain, research which traced the effects of nicotine more specifically was developed primarily in two locations. First, at the Institute of Psychiatry in the mid-1970s, where the technical development of the blood nicotine assay helped to establish the role of nicotine as the major controlling factor in smokers' regulation of smoke intake.[26] Secondly, the tobacco industry was also establishing a sophisticated understanding of the role of nicotine in smoking behaviour. In the 1960s at the industry funded Tobacco Research Council's laboratories in Harrogate, researchers were investigating the role of nicotine in habituation through animal experiments, as part of the industry's focus post-1962 on developing safer smoking. This work was published in *Nature*,[27] and other reports of the time show the clear industrial pharmacological interest in this aspect of smoking research.[28]

However, the developing concept of pharmacological addiction had little social and policy impact in the 1970s. The main 'treatment' model was smoking cessation, based on psychological models of behaviour modification and smoking as socially learned behaviour. Psychologists with an interest in decision making processes used research in anti-smoking clinics as part of this more general psychological emphasis. The primary 'treatment focused' concept in common use was psychologically-based dependence, rather than pharmacological addiction. The mainstream public health emphasis within anti-smoking activities was far removed from pharmacological research networks. The public health emphasis on education, on stopping smoking, on personal responsibility and free will was at odds with some of the implications of nicotine addiction. In the British context, there was no strong science/policy community around the concept of nicotine addiction in this decade. Public health and psychological perspectives predominated.[29]

The 1980s saw rapid change. Nicotine research expanded rapidly, in particular in the US after the earlier start in Britain in the 1960s and 1970s. Animal self-administration studies, together with many other types of study – neurochemical, absorption and dependence, craving and withdrawal, titration and effectiveness of nicotine replacement therapy (NRT) – all showed that nicotine was addictive. Behavioural studies in both Britain and the US finally confirmed that withdrawal symptoms were nicotine related. NRT was found to reduce withdrawal symptoms.

These studies of nicotine were part of the huge expansion of psychopharmacology in this decade. Biomedical work was developed in alliance with epidemiological models and with biomarkers.[30,31] Nicotine research emerged in the US in policy terms in the Surgeon-General's report on nicotine addiction in 1988.[32] Like passive smoking (the other new 'scientific fact') of the 1980s, the acceptance of nicotine addiction had major effects on policy discussion in the 1990s. In the more legalistic context of US smoking policy, the concept became important in liability law suits involving the tobacco industry, in moves to regulate nicotine as a drug by the Food and Drug Administration (FDA) (see Chapter 8).

In the British context, the policy response to addiction has been different, with a revived focus on harm reduction, which had earlier been of significance in the late 1970s, but with relatively little emphasis on the control of either supply or use of nicotine products. The role of nicotine as the agent of tobacco addiction raised significant policy issues. It was pointed out that people smoke mainly for nicotine, but die from the tar and other unwanted components in the smoke.[33] The establishment of the concept of nicotine addiction has dual implications for policy and practice. On the one hand, it could be a reason to maintain people on the drug; on the other, it could be a reason to wean them off the habit.[31] There is a greater emphasis on the role of treatment and NRT, and a shift of position in the public health coalition to accommodate harm reduction.[34]

The contrast between the British and American responses to nicotine addiction recalls the different policy positions on illicit drugs in the 1920s, Britain adopting a medical maintenance model while the US favoured prohibition.

In general, these debates and the associated rise of the concept of addiction can be seen as part of the repositioning of both tobacco and nicotine, and of illicit drug use in the late 20th century. They epitomise the biomedical reorientation of late 20th century public health.[29] However, it is evident that throughout this century nicotine products have tended to enjoy relatively favoured status compared with other addictive drugs, and that this status is starkly inconsistent with the harm that tobacco products cause.

4.4 How does nicotine addiction compare with addiction to other drugs?

In determining the appropriate medical, social and policy responses to nicotine addiction in Britain, it is appropriate to try to position nicotine in relation to other addictive drugs in terms of the degree of that addiction or dependence. In addressing this issue, however, it is important also to acknowledge the importance of drug delivery methods in determining dependence. This point was summarised in testimony by Dr Louis Harris before the US FDA on 3rd August 1994, on behalf of the College on Problems of Drug Dependence and the American Society of Pharmacology and Experimental Therapeutics. He stated that the development of dependence or addiction to drugs such as nicotine may depend on the method by which the drug is delivered, as well as the inherent characteristics of the drug itself, and testified as follows:

> First, the great preponderance of data from both animals and man
> indicate that nicotine meets the criteria to be classified as an abusable

and dependence producing substance. Thus, nicotine produces tolerance and dependence such that abstinence after appropriate dosing may result in withdrawal symptoms. In addition, the compound produces alterations of mood in humans and serves as a reinforcer and is self-administered by both animals and man. That's a given. Second, the psychoactive effects of nicotine are dependent on both dose and rate of administration and route of administration as well as rate. The inhalation route can provide high doses at a rapid rate that produce and sustain dependence.

Similar conclusions were drawn by the US Surgeon-General in 1988[32] and by the FDA.[35,36] The fact that several characteristics of addictive drugs might be scientifically evaluated, and also that the effects of the drug depend in part on how it is delivered, complicate simple statements that equate all forms of nicotine delivery. These considerations also complicate comparisons of addictiveness of nicotine and other drugs because the answer might depend upon the measure and dosage form under consideration. None the less, it is important to try to evaluate the extent to which addiction to nicotine compares with addiction to other drugs. Is nicotine among the most addictive drugs of all, or is it no more addictive than coffee, tea, 'twinkies' (an American confectionery), jogging or carrots – as claimed by the tobacco industry in its statements before the US Congress (see Refs 36 and 37)? The answer to this question is complicated by consideration of the specific criteria considered and the dosage form evaluated.

Nicotine dosage delivery forms

First, let us consider the dosage form issue. Tobacco-delivered nicotine maximises the addictive effects of nicotine in several ways:

1 Tobacco products, in general, and cigarettes in particular, provide remarkably palatable dosing systems. These mask the noxious effects of nicotine, which include burning sensations in the mouth, oral, nasal and pharyngeal regions, and which can occur at very low dosages when nicotine is given in pure form.

2 Tobacco products provide a means of rapid, responsive and flexible dosage control which is important because the range of doses at which effects are positively reinforcing is narrow. Armitage *et al*[38] described this as 'finger tip' dosage control with regard to the cigarette. Speed of delivery of nicotine from cigarettes is remarkable, providing nicotine 'hits' to the brain within seconds of smoke delivery to the lung.

3 The engineering of cigarettes and many smokeless tobacco products employs pharmaceutical techniques to control nicotine

dosage so that it will be optimal for the target population. In the case of moist snuff products, for example, products marketed to young people (termed 'starter products' by the company) use small amounts of buffering to produce low pH, slow-release nicotine systems. In contrast, products marketed to established snuff users are highly buffered to provide much higher and more rapidly delivered nicotine doses according to a scheme termed the 'graduation' process by one tobacco company.[35,36] Cigarettes are remarkably fine-tuned dosing systems, with dosage consistently meeting standards set by regulatory agencies such as the FDA for approved drugs. Furthermore, their physical engineering means that cigarettes marketed most heavily can readily deliver approximately 1–3 mg of nicotine, as desired by smokers, regardless of their advertised or labelled delivery values.[39]

4   The extreme lengths to which tobacco manufacturers went to employ physical and chemical engineering in cigarette design to maximise the addictive effects of nicotine have only recently been appreciated through the discovery by the FDA of documents and collections of expert accounts from former tobacco industry employees.[35,36] The discovery process continues almost daily with reviews of tobacco industry documents now available because of litigation in the US. Taken together, this extraordinary engineering optimises the pH, the ratio of free base to bound nicotine in the smoke, the size of the inspired particles, the sensory effects of the inspired particles, and many other features of modern cigarettes. Cigarette smoke also contains chemicals which can act synergistically to produce effects that might be even more reinforcing than those of nicotine alone. For example, levels of acetaldehyde (a chemical involved in alcohol dependence) in smoke can be manipulated so as to produce a mixture that studies by Philip Morris[35,36] indicated would be more reinforcing than either acetaldehyde or nicotine alone.

5   Finally, other substances in cigarette smoke may produce effects that contribute to the addictive effects of cigarettes. Although research in this area is in its infancy, it is plausible that effects such as the potentially antidepressant monoamine oxidase inhibiting effects of smoke are due to non-nicotine compounds in tobacco and/or tobacco smoke.[40]

Whether the intent of this engineering is to maximise or minimise abuse potential, many aspects of the product itself, and how it is regulated and marketed operate to determine overall patterns of use (see Table 4.2).[41] In contrast to tobacco products, approved

Table 4.2. Comparison of product design, labelling, and marketing strategies and controls for tobacco products and nicotine treatment preparations (from Ref 41).

	Tobacco products	Treatment products
Product design	Maximise pleasurable/ reinforcing effects of addiction	Minimise all addictive effects
	Maximise sensory appeal	Target acceptability to indicated consumer
	Maximise packaging appeal	Target appeal to indicated consumers
	Health risks 'accepted' and conferred to consumers	Health risks minimised 'Safety standards' set by regulation
Labelling	Covered by EU directive Not subject to MCA approval Health warnings required to cover 4–6% of pack surface	MCA approval required before marketing Specific indications, cautions, warnings and messages required
	Requirement to display machine-measured tar and nicotine yields Human dose exposure data not required	Accurate data on nicotine dose content and delivery required
	Exposure reduction claims (eg 'light', 'lower', 'reduced') unregulated	All claims relating to health improvement subject to MCA approval
Marketing	Initiate use in non-tobacco users	Use only for tobacco users
	Create nicotine dependence	Treat nicotine dependence
	Sustain use as long as possible	Use no longer than labelling or medical need recommends
	Explicit youth targeting of some products	Minimise youth appeal

EU = European Union; MCA = Medicines Control Agency.

nicotine-delivering drug products for treating tobacco dependence and withdrawal provide pure forms of nicotine in which dosage is controlled to minimise adverse (including addictive) effects. For example, nicotine patches deliver their nicotine so slowly as to preclude psychoactive effects. Delivery of nicotine from gum is potentially faster but heavily dependent upon determined activity by the user and still cannot mimic the cigarette. The sensory palatability of the pharmaceuticals is also finely balanced so as not unduly to reduce compliance with clinical guidance but also not to be overly attractive in their own right. Consequently, the users of nicotine products such as oral gum, inhalers and nasal spray obtain little pleasure from use of these products, and in fact experience sensory stimuli which tend to discourage unnecessary use. These differences are best illustrated by the fact that, whereas most smokers use cigarettes much more and for much longer than they desire, most users of the nicotine replacement medicines use doses at levels below those advised by health professionals, and the incidence of abuse of these products is remarkably low.

Nicotine dosage delivery by cigarette

Keeping in mind the tremendous variation in the addictive effects of nicotine across delivery systems, let us now focus on the form which is the most troublesome from a health perspective, namely, the cigarette. Sufficient epidemiological and clinical evidence enables a systematic comparison of cigarettes and other addictive drugs according to various criteria. One such comparison was provided in a table in the 1988 Surgeon-General's report.[32]

More recently, Henningfield et al[42] compiled a similar comparison (see Table 4.3). In brief, as shown in the table, the severity of the addictive effects varies across different measures. Although others may rate these features somewhat differently,[43] the point remains that several different features of drugs might be considered and that no addictive drug exceeds all others on all points. As illustrated in Table 4.3, and as has been concluded elsewhere,[32,44,45] among the prototypic addictive drugs, nicotine delivered by way of tobacco

Table 4.3. Ranking of nicotine in relation to other drugs in terms of addiction factors of concern (from Ref 42).

Dependence among users	nicotine>heroin>cocaine>alcohol>caffeine
Difficulty achieving abstinence	(alcohol=cocaine=heroin=nicotine)>caffeine
Tolerance	(alcohol=heroin=nicotine)>cocaine>caffeine
Physical withdrawal severity	alcohol>heroin>nicotine>cocaine>caffeine
Societal impact	serious effects due to secondary deaths (nicotine), accidents (alcohol) or crime (heroin, cocaine); no substantial impact for caffeine
Deaths	nicotine>alcohol>(cocaine=heroin)>caffeine
Importance in user's daily life	(alcohol=cocaine=heroin=nicotine)>caffeine
Intoxication	alcohol>(cocaine=heroin)>caffeine>nicotine
Animal self-administration	cocaine>heroin>(alcohol=nicotine)>caffeine
Liking by non-drug abusers	cocaine>(alcohol=caffeine=heroin=nicotine)
Prevalence	caffeine>nicotine>alcohol>(cocaine=heroin)

products is a highly addictive drug. None the less, differences in specific features of addictive drugs and differences in the consequences of use have implications for regulatory approaches appropriate to each drug as well as to clinical approaches to treating individuals determined by the particular drug under consideration.[45] Prominent features of particular distinction across addictive drugs are summarised below.

Incidence, prevalence and risk of progression

Addiction to nicotine is far more common than addiction to cocaine, heroin or alcohol, and the rate of graduation from occasional use to addictive levels of intake is highest for nicotine in the form of cigarettes. Depending upon the definition used for occasional use, 33–90% of occasional users escalate to become daily smokers[46] (see also Chapter 5). In contrast, even when highly addictive dosage forms of cocaine (ie smokeable 'crack' cocaine) are readily available in the US, the risk of progression from any use to regular use is the exception, not the rule. The 1988 US National Household Survey indicated that cocaine is currently used at least once per week by about 11% of people who have used cocaine in the past year and by about 29% of people who have used it in the past 30 days. For alcohol, approximately 10–15% of consumers of alcoholic beverages are problem drinkers. Although the absolute estimates may vary, an epidemiological study by Anthony *et al*[47] on the risk of dependence according to the DSM criteria confirms these comparisons and conclusions.

Remission and relapse

Rates and patterns of relapse are similar for nicotine, heroin and alcohol,[48,49] and probably for cocaine.[50] An analysis of relapse to tobacco use showed that, in the context of a minimal treatment intervention approach, approximately 25% of persons relapsed within two days of their last cigarette and approximately 50% within one week.[51,52] For people quitting on their own, the study by Hughes *et al*[52] discovered that two-thirds were smoking within three days of their scheduled quit date.

Reports of addictiveness by drug abusers

Two studies specifically asked polydrug abusers to compare their addictions. The first asked drug abusers to rate their liking on an increasing scale from 1–4.[53] Tobacco, cocaine, heroin and alcohol liking scales were 4.3, 4.2, 4.7 and 2.9, respectively. On the need scale, tobacco was rated most highly (3.3) and alcohol most weakly (1.3), while heroin was rated at 2.8 and cocaine at 1.5. A second study[54] found that tobacco, when compared to other substances, was associated with equal or greater levels of difficulty in quitting and urge to use, but that its use was not as pleasurable. Using a laboratory-based approach, Henningfield et al[55] found that cigarette smokers who also had histories of other drug abuse rated intravenous (IV) nicotine as similar to cocaine on key measures of addiction potential. These findings were recently extended by Jones et al[56] in a direct, double-blind comparison of nicotine with cocaine given IV to human volunteers. This study also found similar effects of the two drugs on key measures of addiction potential. Of particular note was the finding that subjects frequently misidentified nicotine as cocaine, and at high doses, as an opiate.

Psychoactivity and euphoria

Among the first steps in determining whether a chemical has the potential to produce addiction is to determine if it is psychoactive.[32] Psychoactive effects are often referred to in human studies as subjective, psychological, interoceptive or psychic, or in both animal and human studies as discriminative. As described earlier, nicotine and other comparison drugs all produce qualitatively distinct psychoactive effects. Of course, not all psychoactive drugs are addictive. Drugs such as chlorpromazine or atropine are psychoactive but not widely abused, and the psychoactive effects produced by these drugs are not generally considered highly pleasant in their own right.

One correlate of addiction liability is that a drug produces pleasurable or euphoriant effects in standard tests of drug liking and morphine-benzedrine group (MBG) scale scores.[57–59] In polydrug users, scores on liking scales do not necessarily show quantitative differences between nicotine, cocaine, heroin, or alcohol.[32] Scores on the MBG scale, however, are elevated by most addictive drugs, and these absolute values do vary across drugs.[59] Such variation probably reflects qualitative differences in the effects of the drugs and not quantitative differences in addictiveness.

Reinforcing effects

The capacity of a drug to control behaviour leading to its repeated self-administration can be tested by giving animals or humans the opportunity to take it under standardised conditions. Nicotine, cocaine, heroin and alcohol serve as reinforcers for a variety of species.[32,60] Cocaine appeared to be the more powerful reinforcer in several studies in which nicotine has been directly compared with cocaine.[61-64] Analogous comparisons with opioids and alcohol have not been made, nor have other routes of drug administration been compared, thus weakening the strength of conclusions regarding possible differences in the maximal reinforcing potential of these drugs. None the less, cocaine appears to be the most readily established reinforcer for animals, generally requiring only simple access to the drug via an IV catheter.[65,66] Therefore, whereas such studies confirm that the drug nicotine in tobacco products, the drug morphine in opium products, and the drug cocaine in coca-based products, respectively, define the drug dependency syndromes, such studies do not provide a basis for predicting how the reinforcing effects of the drugs will compare in products used outside the laboratory.

Physical dependence

Many drugs that are not abused also produce physical dependence, (eg anticholinergics, dopaminergic antagonists and calcium channel blockers).[67] Among the addicting drugs, the most severe withdrawal syndromes are those which occur following extended administration of alcohol or short-acting barbiturates.[68] Heroin and nicotine also produce clearly defined syndromes of physical dependence and withdrawal,[32] and a syndrome of withdrawal from chronic cocaine administration has also been recently characterised.[69] The symptoms of withdrawal from cigarettes appear to exceed those for all other forms of nicotine delivery; they are less severe than those produced by alcohol or heroin, but more severe than those from cocaine.[32,70]

Tolerance

The degree and type of tolerance that occur vary considerably across drugs. For example, nicotine, cocaine, heroin and alcohol can produce intoxication and disorientation,[67] but tolerance to the intoxicating effects of nicotine and heroin is sufficiently pronounced for intoxication to be relatively uncommon in users with stable supplies of drugs.[32,68] Conversely, the degree of behavioural tolerance to alcohol is so limited that automobile accidents are common in heavy drinkers. Nicotine tolerance has been widely studied since the turn of the 20th century.[71] The degree of tolerance produced by nicotine is so pronounced that it can enable tobacco users to self-administer the large quantities of tobacco each day that lead to the high risk of disease and premature death associated with cigarette smoking.

Conclusions

On current evidence, we can conclude that cigarettes are properly categorised among the most addicting substances as this form of nicotine delivery maximises the addictive effects of the drug. The fact that nicotine is of low abuse potential in controlled dosage forms such as the transdermal nicotine patch or nicotine gum supports the conclusion that the form of delivery is an important determinant of its addiction potential. Thus, tobacco-delivered nicotine is of great concern, with the cigarette of greatest concern of all tobacco products because of its high toxicity and addictiveness.

The pharmacological effects of nicotine are not identical to those of heroin, alcohol or cocaine – nor, for that matter, are the effects of cocaine identical to those produced by heroin. In its arguments that nicotine is not addictive, the tobacco industry has often argued, as it did to the US FDA,[36] that nicotine is not addicting because it does not meet criteria that the tobacco industry itself has developed. In essence, these criteria appear to be those achievable only by a drug whose composite profile would be as intoxicating as ethanol, would produce as severe withdrawal symptoms as ethanol or heroin, would have the euphoriant effects of cocaine, and would serve as a reinforcer for animals and naïve humans as readily as does cocaine.

Any one factor may be selected to argue that one of these drugs is more or less addicting than the others. However, this exercise makes it clear that addiction severity and society's level of concern about drug use are best evaluated by assessing several variables. We can, however, conclude, as was concluded in the 1988 Report of the US Surgeon-General,[32] that:

> The pharmacologic and behavioral processes that determine tobacco addiction are similar to those that determine addiction to drugs such as heroin and cocaine.

We can further conclude that tobacco dependence is a serious form of drug addiction which, on the whole, is second to no other.

4.5 Relevance to society of recognition of nicotine as an addictive drug

As described in Section 4.3, dependence on smoking and nicotine addiction has historically been afforded a relatively lower degree of significance than other forms of drug addiction in Britain. Indeed, it is still widely claimed that tobacco use is a pleasurable activity in which adults participate knowing and accepting the risks and, according to the Tobacco Manufacturers' Association:

Smoking is an adult pursuit and should remain a matter for informed and adult choice.[72]

Further, the principle of consumer sovereignty is well established in modern economic theory, and holds that consumers are the best judges of how to spend their own money. However, if smoking and nicotine are addictive, the argument that the individual adult consumer has a right to choose to purchase and use tobacco products, and that the tobacco industry has a right to continue to supply them, is difficult to sustain for the following reasons:

1   Addiction to tobacco products creates a demand for habitual use which is no longer an expression of consumer choice. This point was acknowledged in confidential legal advice to the US Tobacco Institute in 1980, that:

> we can't defend continued smoking as 'free choice' if the person was 'addicted'.[73]

The addictiveness of tobacco products compromises the free choice and consumer sovereignty of the smoker. Although it is possible to stop smoking, many find smoking cessation a difficult and unexpected consequence of smoking, and for some it represents an insurmountable obstacle. Deciding to smoke or make a purchase of cigarettes has an element of future commitment to smoking that may not be apparent to the smoker at the time of purchase.

2   The vast majority (over 80%) of UK smokers start as teenagers or children.[74] Young people do not always have the capacity to make informed decisions, and society generally recognises this by providing greater protections for children than for adults. For example, there are a large number of age-restricted products on the market including tobacco, and special children's rights are enshrined in a UN convention with articles establishing the child's right to good health.[75] In the case of tobacco, the problem of immature decision making is compounded by nicotine addiction. Decisions to smoke made in the early teens can be consolidated into addictive behaviour before the smoker reaches maturity. A confidential Philip Morris marketing document explains this as follows:

> a cigarette for the beginner is a symbolic act. I am no longer my mother's child, I'm tough, I am an adventurer, I'm not square ... As the force from the psychological symbolism subsides, the pharmacological effect takes over to sustain the habit.[76]

3   Consumers can have poor awareness of the risks, and therefore make suboptimal decisions. This type of 'imperfection' is common

in many product markets, but the consequences of ill-informed decisions about tobacco products may lead to serious harm or premature death. Though smokers generally know they face increased risks and are warned to that effect, they judge the magnitude and diversity of these risks to be lower and less well established than they actually are.[77] As well as poor knowledge of the health risks such as cancer and heart disease, there is also poor knowledge of the risk of addiction, compounded by the difficulty in conceptualising addictive behaviour before experiencing it. A majority of school-age regular smokers believe they are no less or no more likely than others to be 'hooked on cigarettes when 20'.[78] In the US, a study showed that teenagers overestimate the likelihood that they will quit smoking in five years' time.[79] This knowledge deficit is not merely consumer behaviour that falls short of a theoretical ideal, but is at least in part the outcome of sustained public relations campaigns by the tobacco industry. For example, British American Tobacco advertised in *The Observer*, a British Sunday newspaper, in March 1998 stating:

> Whereas earlier definitions [of addiction] were based on objective criteria, the current definition is more colloquial, reflected in terms like 'chocaholic' and 'Addicted to Love' as in a recent movie. This colloquial definition is all inclusive and certainly applies to the use of many common substances that have similar mild pharmacological effects to cigarettes, such as coffee, tea, chocolate and cola drinks.[80]

The fact is, however, that in relation to the more objective criteria, smoking and nicotine use are addictive. They differ significantly from the colloquially defined addictions listed above in both the severity and strength of that addiction, and in the adverse health effects caused by the addiction. For practical purposes, and in major contrast to cigarettes, chocolate, coffee, tea, cola drinks and love do not kill.

References

1   American Psychiatric Association. Position statement on nicotine dependence. *Am J Psychiatry* 1995; **152**: 481–2.
2   World Health Organization. *International Statistical Classification of Diseases and related Health Problems*, 10th revision. Geneva: WHO, 1992.
3   World Health Organization. Nomenclature and classification of drug- and alcohol-related problems: a WHO memorandum. *Bull WHO* 1964; **90**: 225–42.
4   West R, Gossop M. A comparison of withdrawal symptoms from different drug classes. *Addiction* 1994; **89**: 1483–9.
5   Jaffe J. Tobacco smoking and nicotine dependence. In: Wonnacott S, Russell MAH, Stolerman IP (eds). *Nicotine psychopharmacology: molecular, cellular and behavioural aspects*. Oxford: Oxford University Press, 1990: 1–37.

6   American Psychiatric Association. *Diagnostic and Statistical Manual of Mental Disorders*, 3rd edn (revised). Washington: APA, 1987.

7   American Psychiatric Association. *Diagnostic and Statistical Manual of Mental Disorders*, 4th edn. Washington: APA, 1995.

8   Freeth S. *Smoking-related behaviour and attitudes, 1997*. A report on research using the Omnibus Survey produced on behalf of the Department of Health. London: Office for National Statistics, 1998.

9   Bridgewood A, Malbon G, Lader D, Matheson J. *Health in England 1995*. London: The Stationery Office, 1996.

10 West R, McEwen A, Bates C. *Sex and smoking*. London: No Smoking Day, 1999.

11 Jarvis MJ. Patterns and predictors of smoking cessation in the general population. In: Bolliger CT, Fagerström KO (eds). *The tobacco epidemic. Progress in Respiratory Research* 1997; **28**: 151–64. Basel: Karger.

12 Johnson EO, Breslau N, Anthony JC. The latent dimensionality of DIS/ DSM-IIIR nicotine dependence: exploratory analyses. *Addiction* 1996; **91**: 583–8.

13 Levine H. The discovery of addiction: changing conceptions of habitual drunkenness in America. *J Stud Alcohol* 1978; **39**: 143–74.

14 Porter R. The drinking man's disease: the pre-history of alcoholism in Georgian Britain. *Br J Addict* 1985; **80**: 385–96.

15 Berridge V. *Opium and the people. Opiate use and drug control policy in nineteenth and early twentieth century England* (expanded edn). London: Free Association Books, 1998.

16 Thom B. *Dealing with drink. Alcohol and social policy: from treatment to management*. London: Free Association Books, 1999.

17 Kerr N. *Inebriety*. London: H.K. Lewis, 1888.

18 Welshman J. Images of youth: the problem of juvenile smoking, 1900–1939. *Addiction* 1996; **91**: 1379–86.

19 Hilton M. *Consumer society in England, 1850–1950, with special reference to tobacco*. PhD thesis. Lancaster: University of Lancaster, 1996.

20 Dixon WE. The tobacco habit. *Br J Inebriety* 1927–28; **25**: 99–121.

21 Berridge V. Science and policy: the case of post war British smoking policy. In: Lock S, Reynolds L, Tansey EM (eds). *Ashes to ashes. The history of tobacco and health*. Amsterdam: Rodopi, 1998.

22 Lewis J. *What price community medicine? The philosophy, practice and politics of public health since 1919*. Brighton: Wheatsheaf, 1986.

23 Johnston LM. Tobacco smoking and nicotine. *Lancet* 1942; **ii**: 742.

24 Royal College of Physicians. *Smoking and health*. London: Pitman Medical Publishing Co, 1962.

25 Royal College of Physicians. *Smoking and health now*. London: Pitman Medical Publishing Co, 1972.

26 Russell MAH. Tobacco smoking and nicotine dependence. In: Gibbins RJ, Israel Y, Kalant H, Popham RE, *et al* (eds). *Research advances in alcohol and drug problems*. New York: John Wiley and Sons, 1976: 1–47.

27 Armitage AK, Hall GH, Morrison KF. Pharmacological basis for the tobacco smoking habit. *Nature* 1968; **217**: 331–4.

28 von Euler US. *Tobacco alkaloids and related compounds*. Oxford: Pergamon Press, 1965.

29 Berridge V. Changing places: illicit drugs, medicines, tobacco and nicotine in the nineteenth and twentieth centuries. In: Tansey EM, Gijswit-Hofstra M (eds). *Remedies and healing cultures in Britain and the Netherlands in the twentieth century* (in press).

30 CIBA symposium. *The biology of nicotine dependence*. Chichester: John Wiley and Sons, 1990.

31   Wald N, Froggatt P (eds). *Nicotine, smoking and the low tar programme.* Oxford: Oxford University Press, 1989.

32   US Department of Health and Human Services. *The health consequences of smoking, nicotine addiction.* A report of the Surgeon-General, US Department of Health and Human Services, DHHS Publication No. (CDC) 88-8406. Washington: DHHS, 1988.

33   Anon. Nicotine use after the year 2000. *Lancet* 1991; **337**: 1191–2.

34   Raw M. *Regulating nicotine delivery systems. Harm reduction and the prevention of smoking related disease.* London: Health Education Authority, 1997.

35   Food and Drug Administration. Regulations restricting the sale and distribution of cigarettes and smokeless tobacco products to protect children and adolescents; proposed rule. *Fed Register* 1995; **60**: 41313–787.

36   Food and Drug Administration. Regulations restricting the sale and distribution of cigarettes and smokeless tobacco products to protect children and adolescents; final rule. *Fed Register* 1996; **61**: 44395–5318.

37   Reynolds RJ. *Regulations of tobacco products I. Hearings before the Subcommittee on Health and the Environment of the Committee on Energy and Commerce House of Representatives.* Serial No. 103-149. Washington: US Government Printing Office, 1994: 579.

38   Armitage AK, Hall GH, Morrison CF. Pharmacological basis for the tobacco smoking habit. *Nature* 1968; **217**: 331–4.

39   National Cancer Institute, National Institute of Health. *The FTC cigarette test method for determining tar, nicotine, and carbon monoxide yields of US cigarettes: report of the NCI Expert Committee.* Smoking and Tobacco Control Monograph 7. (NIH Pub. No. 96-4028). Bethesda, MD: NIH, 1996.

40   Fowler JS, Volkow ND, Wang G-J, Pappas N, *et al.* Brain monoamine oxidase A inhibition in cigarette smokers. *Proc Natl Acad Sci USA* 1996; **93**: 14065–9.

41   Warner KE, Peck CC, Woosley RL, Henningfield JE, Slade JS. Treatment of tobacco dependence: innovative regulatory approaches to reduce death and disease. Preface. *Food Drug Law J* 1998; **53**: 2.

42   Henningfield JE, Schuh LM, Heishman SJ. Pharmacological determinants of cigarette smoking. In: Clarke PBS, Quik M, Adlkofer FX, Thurau K (eds). *Effects of nicotine on biological systems II.* International Symposium on Nicotine Advances in Pharmacological Sciences. Basel: Birkhauser Verlag, 1995: 254.

43   Hilts PJ. Is nicotine addictive? It depends on whose criteria you use. *The New York Times*, 2nd August 1994, p. C3.

44   Henningfield JE, Cohen C, Slade JD. Is nicotine more addictive than cocaine? *Br J Addict* 1991; **86**: 565–9.

45   Goldstein A. *Addiction: from biology to drug policy.* New York: W.H. Freeman and Co, 1994.

46   US Department of Health and Human Services. *Preventing tobacco use among young people: a report of the Surgeon-General.* Centers for Disease Control and Prevention, National Center for Chronic Disease Prevention and Health Promotion, Office on Smoking and Health. Atlanta, GA: USDHHS, 1994.

47   Anthony JC, Warner LA, Kessler RC. Comparative epidemiology of dependence on tobacco, alcohol, controlled substances, and inhalants: basic findings from National Comorbidity Survey. *Exp Clin Psychopharmacol* 1994; **3**: 244–68.

48   Hunt WA, Matarazzo JD. Three years later: recent developments in the experimental modification of smoking behavior. *J Abn Psychol* 1973; **31**: 107–14.

49   Maddux JF, Desmond DP. Relapse and recovery in substance abuse careers. In: Tims FM, Leukefeld CG (eds). *Relapse and recovery in drug abuse.* National Institute of Drug Abuse Research Monograph 72. US Department of Health

and Human Services, Public Health Service, Alcohol, Drug Abuse, and Mental Health Administration, NIDA. DHHS Publication No. (ADM) 88-1473. Rockville: USDHHS,1986: 49–71.

50 Wallace BC. Psychological and environmental determinants of relapse in crack cocaine smokers, *J Subst Abuse Treat* 1989; **6**: 95–106.

51 Kottke TE, Brekke ML, Solberg LI, Hughes JR. A randomized trial to increase smoking intervention by physicians, doctors helping smokers. Round 1. *JAMA* 1989; **261**: 2101–6.

52 Hughes JR, Gulliver SB, Fenwick JW, Valliere WA, *et al.* Smoking cessation among self-quitters. *Health Psychol* 1992; **11**: 331–4.

53 Blumberg HH, Cohen SD, Dronfield BE, Mordecai EA, *et al.* British opiate users I. People approaching London drug treatment centres. *Int J Addict* 1974; **9**: 1–23.

54 Kozlowski LT, Wilkenson DA, Skinner W, Kent C, *et al.* Comparing tobacco cigarette dependence with other drug dependencies. *JAMA* 1989; **261**: 898–901.

55 Henningfield JE, Miyasato K, Jasinski DR. Abuse liability and pharmaco-dynamic characteristics of intravenous and inhaled nicotine. *J Pharmacol Exp Ther* 1985; **234**: 1–12.

56 Jones HE, Garrett BE, Griffiths RR. Subjective and physiological effects of intravenous nicotine and cocaine in cigarette smoking cocaine abusers. *J Pharmacol Exp Ther* 1999; **288**: 188–97.

57 Martin WR. *Drug addiction I. Handbook of Experimental Pharmacology*, vol. 45/1. Heidelberg: Springer-Verlag, 1977: 197–258.

58 Fischman MW, Mello NK (eds). *Testing for abuse liability of drugs in humans.* National Institute of Drug Abuse Research Monograph 92. US Department of Health and Human Services, Public Health Service, Alcohol, Drug Abuse, and Mental Health Administration, NIDA. DHHS Publication No. (ADM) 89–1613. Rockville: USDHHS, 1989.

59 Jasinski DR, Henningfield JE. Human abuse liability assessment by measure-ment of subjective and physiological effects. In: Fischman MW, Mello NK (eds). *Testing for abuse liability of drugs in humans.* National Institute of Drug Abuse Research Monograph 92. US Department of Health and Human Ser-vices, Public Health Service, Alcohol, Drug Abuse, and Mental Health Administration, NIDA. DHHS Publication No. (ADM) 89-1613. Rockville: USDHHS, 1989: 73–100.

60 Corrigall WA. Nicotine self-administration in animals as a dependence model. *Nicotine Tob Res* 1999; **1**: 11–20.

61 Ator NA, Griffiths RR. Intravenous self-administration of nicotine in the baboon. *Fed Proc* 1980; **40**: 298.

62 Goldberg SR, Spealman RD. Maintenance and suppression of responding by intravenous nicotine injections in squirrel monkeys. *Fed Proc* 1982; **421**: 216–20.

63 Griffiths RR, Brady V, Bradford LD. Predicting the abuse liability of drugs with animal drug self-administration procedures: psychomotor stimulants and hallucinogens. In: Thompson T, Dews PB (eds). *Advances in behavioral pharmacology*, vol. 2. New York: Academic Press, 1979: 163–208.

64 Risner ME, Goldberg SR. A comparison of nicotine and progressive-ratio schedules of intravenous drug infusion, *J Pharmacol Exp Ther* 1983; **224**: 319–26.

65 Deneau G, Yanagita T, Seevers MH. Self-administration of psychoactive substances by the monkey. A measure of psychological dependence. *Psychopharmacologia* 1969; **16**: 30–48.

66 Pickens R, Thompson T. Cocaine-reinforced behavior in rats. Effects of rein-forcement magnitude and fixed-ratio size. *J Pharmacol Exp Ther* 1968; **161**: 122–9.

67 Gilman AG, Rall TW, Nies AS, Taylor P (eds). *Goodman and Gilman's the pharmacological basis of therapeutics*, 8th edn. New York: Macmillan, 1990.

68 Jaffe JH. Drug addiction and drug abuse. In: Gilman AG, Rall TW, Nies AS, Taylor P (eds). *Goodman and Gilman's the pharmacological basis of therapeutics*, 8th edn. New York: Macmillan, 1990: 522–73.

69 Gawin FH, Kleber HD. Abstinence symptomatology and psychiatric diagnosis in cocaine abusers. *Arch Gen Psychiatry* 1986; **43**: 107–13.

70 O'Brien CP. Drug addiction and drug abuse. In: Molinoff PB, Ruddon RW (eds). *Goodman and Gilman's the pharmacological basis of therapeutics*, 9th edn. New York: McGraw-Hill, 1996: 557–77.

71 Swedberg MDB, Henningfield JE, Goldberg SR. Nicotine dependency: animal studies. In: Wonnacott S, Russell MAH, Stolerman IP (eds). *Nicotine psychopharmacology: molecular, cellular, and behavioural aspects*. Oxford: Oxford University Press, 1990: 38–76.

72 Tobacco Manufacturers' Association (UK). *Smoking and children*, 1997. http://www.thetma.org.uk/ children and smoking.htm.

73 Knopick P. US Tobacco Institute. Memo to W Kloepfer, September 1980. Minnesota Trial Exhibit 14,303. http://www.mnbluecrosstobacco.com/ toblit/trialnews/docs/TE14,303.pdf.

74 Thomas M, Walker A, Wilmot A, Bennett N. *Living in Britain: results from the 1996 General Household Survey*. Office for National Statistics. London: The Stationery Office, 1998.

75 *UN Convention on the Rights of the Child*. Articles 6 and 24. UNICEF, 1989. http://www.unicef.org/crc/fulltext.htm.

76 Dunn W. Vice President for Research and Development, Philip Norris. *Why one smokes*. Minnesota Trial Exhibit 3681. http://www.mnbluecrosstobacco.com/ toblit/trialnews/docs/TE3681.pdf.

77 World Bank. *Curbing the epidemic: governments and the economic of tobacco control*. Washington: World Bank Publications, June 1999.

78 Jarvis L. Office for National Statistics. *Teenage smoking attitudes in 1996* (Table 4.13). London: The Stationery Office, 1997.

79 Centers for Disease Control and Prevention. *Preventing tobacco use among young people: a report of the Surgeon-General*. Washington: US Department of Health and Human Services, 1994.

80 British American Tobacco (advertisement). *The Observer*, London. 1st March 1998.

5 | The natural history of smoking: the smoker's career

5.1 Nicotine intake in novice smokers, and the development of dependence

5.2 Persistent and compulsive smoking in the face of smoking related ill-health

5.3 Non-dependent smokers

5.4 Smoking cessation rates in Britain

5.1 Nicotine intake in novice smokers, and the development of dependence

The vast majority of people who become regular smokers begin their smoking career in adolescence. It is currently difficult to assess dependence on nicotine at this age because the available standard instruments and criteria for measuring and determining dependence have been developed primarily for adults. They do not necessarily therefore take account of the special constraints on smoking that apply to young people, such as not being able to smoke at home or at school, or of the financial and legal limitations on the availability of cigarettes at this age. There are as yet no agreed specific criteria for defining and diagnosing nicotine dependence in young smokers. Some preliminary research has been carried out with older teenagers in the US and New Zealand (summarised below). In addition, small-scale studies and national surveys of young people's smoking in England have sometimes included saliva cotinine measurements; the findings of these in relation to the development of dependence are also discussed.

Measuring nicotine dependence in adolescents

In the US, Prokhorov *et al*[1] modified the Fagerström Tolerance Questionnaire (FTQ) for use in a group of over 100 vocational technical students aged 15–20 years (an atypical sample shown to be more likely to start smoking earlier, and to smoke more heavily than their high school counterparts). Although there were marked differences in cigarette consumption and duration of smoking relative to adult populations, the overall FTQ was close to that of a sample of highly nicotine-dependent adults. The FTQ values were generally lower in the student sample, but 20% had an overall FTQ score of 6 and above (compared with 49% in the adult sample), consistent with substantial nicotine dependence.

In New Zealand, using the Diagnostic and Statistical Manual of Mental Disorders (DSM)-IIIR classification of dependence,[2] 20% of a sample of 900 18 year olds were found to be dependent on tobacco. Among those who had smoked every day for a month in the last year, 56% were classified as tobacco dependent.

Inhalation and nicotine intake

Several UK studies have demonstrated that evidence of smoke inhalation and nicotine intake is demonstrable from saliva cotinine concentrations from a very early stage of smoking.[3–5] In a study of 11–15 year old girls in London,[3] cotinine concentrations in occasional smokers (non-daily smokers) indicated that some were already inhaling and obtaining pharmacologically significant doses of nicotine from their cigarettes. By the time a girl was smoking on a daily basis, she appeared to be inhaling a similar dose of nicotine per cigarette to that inhaled by adult smokers.

The Health Survey for England measured salivary cotinine in over 10,000 4–24 year olds in 1996 and 1997. Analyses of these data[5] also demonstrate that inhalation is established even in non-daily smokers. Almost all those claiming to smoke more than six cigarettes a week had cotinine values of 15 ng/ml or above, which is strongly indicative of smoking.[6] For those claiming to smoke between one and six cigarettes a week, 82% of boys and 84% of girls had cotinine values above this level. Cotinine is sensitive to recency of smoking, so the authors concluded that it was likely that the saliva sample was taken after a gap of several days since the last cigarette in the rest of the group. Cotinine levels were higher in the lower income groups, indicating greater exposure to smoke constituents and hence risk of smoking related disease being established in these groups at a young age.

Cotinine levels among smokers increase with age. Given the early establishment of inhalation, it seems likely that these increases are largely due to increases in cigarette consumption rather than to increases in the amount of smoke inhaled per cigarette.[3]

Quit attempts and withdrawal symptoms

The development of inhalation means that nicotine can play an active role in reinforcing smoking from very early in the smoker's career. Consistent with this, studies have also shown that the majority of young smokers perceive themselves to be dependent on smoking. In the study of London schoolgirls,[7] even within their first year of smoking most reported wanting to stop, having tried to do so, and suffering aversive effects when doing so. Daily smokers were more likely to report withdrawal effects than non-daily smokers (74% vs 47%). A withdrawal score measured in this study correlated significantly with saliva cotinine levels and weekly cigarette consumption. It seems likely that withdrawal effects would be partly responsible for the failure of attempts to stop.

A recent national survey of teenagers and smoking in 1997 in England[8] supported these findings. Nearly half (49%) of the current smokers said they would like to give up smoking, and 71% of them reported that they had tried to quit. Nearly half (49%) of regular smokers (defined as usually smoking one or more cigarettes a week) also said they would find it difficult to go without smoking for a day, and 22% that they would find it very difficult. Only 4% of occasional smokers (defined as less than one cigarette a week) said they would find it difficult. Conversely, only 20% of regular smokers said they would find it very easy not to smoke for a day, compared with almost three-quarters (73%) of occasional smokers. The authors concluded that although, on average, children smoke fewer cigarettes than adults, the relative proportions of the regular smokers reporting different degrees of perceived difficulty were comparable to those reported in surveys of adult smokers.

Similar findings have been reported elsewhere.[9] In a US survey of 12–18 year olds[10] the likelihood of reporting symptoms of nicotine withdrawal increased in relation to frequency and intensity of cigarette smoking. Younger and older smokers were equally likely to report increasing nicotine withdrawal symptoms as smoking frequency increased.

Sargent et al[11] examined smoking cessation in 12–18 year old adolescent smokers in the US. They found that smoking cessation rates were 46% among occasional smokers, 12% among daily smokers of 1–9 cigarettes, and 7% of those smoking 10 or more cigarettes per day. The authors reported that this latter rate was comparable with that observed in addicted adult smokers. Although intent to quit smoking was a reliable predictor of cessation among occasional smokers, this was not the case among the adolescent daily smokers, suggesting that nicotine addiction had become important in maintaining cigarette smoking in smokers in this age group.

It is not yet clear how relapse in adolescents compares with adults,[11] or even whether the concept of relapse is applicable to teenagers. Some studies found relapse rates of a lower order among adolescent than among adult quitters, whereas other studies found adolescent quitters to be more likely than adults to relapse. This is an area requiring further research. The difficulty experienced by many adolescent smokers in giving up smoking demonstrates the need for cessation support for this age group.

Time to the first cigarette of the day

How soon people smoke their first cigarette after waking is another measure of addiction. In the 1997 English survey of adolescents,[8] just under half (46%) of the regular teenage smokers usually had their first cigarette within an hour of waking up in the morning, boys being more likely than girls to do so (53% vs 41%). Barton concluded that this was fewer than the percentage of adult smokers who smoke within an hour of waking (64%) (although still high), and could be accounted for by the fact that many children and adolescents cannot smoke at home. Only 13% of regular adolescent smokers usually smoked their first cigarette within 15 minutes of waking, compared with a third of adults.

Subjective effects

Subjective effects of smoking are also commonly reported by adolescent smokers,[8,12] and there is some evidence of consistency in reports of these effects over time.[13] In the British survey,[8] 77% of current smokers said they felt calmer after smoking, 28% felt dizzy, 23% felt more alert and 6% felt sick. Regular smokers were more likely than occasional smokers to feel calmer after smoking (84% vs 62%) and less likely to feel dizzy and sick.

The nature of any positive reinforcement that children get from their smoking is unclear. In the London study,[12] less than a quarter reported a nice feeling from their smoking, suggesting that such positive pleasant feelings were not critical to their smoking. Again, the most prominent subjective effect was feeling 'calmer', and there was a significant correlation between this and saliva cotinine concentrations. This feeling was also related to reports of aversive withdrawal during attempts to stop. Rather than a direct effect of their smoking, feeling calmer may therefore come about as a relief of incipient withdrawal symptoms. The relationship between feeling calmer when smoking and reports of aversive effects when attempting to give up was evident even among those in their first year of smoking, suggesting a rapid development of dependence. There was also some evidence that these effects became more pronounced with more prolonged exposure to smoking.

Gender differences in dependence

Goddard[4] investigated whether sex had an effect on dependence, independent of amount smoked. She developed a composite model of dependence based on questions about craving, difficulty going without smoking, perceived difficulty in giving up smoking, and likely success in quitting. Usual cigarette consumption was the strongest predictor of dependence, with cotinine levels also having a significant effect. Sex was much less important than consumption in predicting dependence, but the analysis indicated that girls appeared more likely to be dependent on cigarettes than boys at a given level of consumption.

Initial smoking experiences

It is known that tolerance can begin with the first dose of nicotine.[14] Very few studies have examined the effects of the first few cigarettes;[15,16] this is an area requiring further research. It is thought that the interpretation of the effects of the first cigarette may be important.[17]

Progression from experimental to regular smoking

Experimental smokers are highly likely to become regular smokers,[18] but exactly what proportion goes on to regular smoking and what factors influence this progression remain largely unknown. It has been argued[19,20] that smoking only a few cigarettes leads to a near inevitable escalation to regular smoking. Other more recent studies have indicated that perhaps one-third to one-half of those young people who experiment with cigarettes go on to become regular smokers.[13] Studies do indicate considerable movement into and out of smoking behaviour over the first few years of smoking. One study, however, found that daily smoking was particularly stable over a two-year period, since 97% of daily smokers in the first survey were still smoking two years later.[13]

Conclusions

All the above studies support an early development of dependence on nicotine in young smokers. Daily smokers appear to be inhaling similar doses of nicotine per cigarette to those inhaled by adults, and are highly likely to want to quit, to have tried to quit, and suffered withdrawal effects when doing so. However, even in many more occasional smokers, inhalation is apparent and there are perceived difficulties in quitting. What can we therefore conclude about the onset of nicotine dependence and models of the development of smoking behaviour?

The development of a regular smoking pattern had been reported to take about two years and was envisaged to involve a progression through a series of stages. Flay *et al*[21] postulated a four-stage model:

- preparatory stage
- trial stage
- experimental stage, and
- regular smoking.

Another model suggested that young smokers may move from initial use, motivated by non-pharmacological factors, to later use, motivated by relief of withdrawal or by other reinforcements due to the pharmacological properties of nicotine.[22] In general, the suggestion is that nicotine gradually takes a stronger hold over time.

Pomerleau *et al*[23,24] have suggested a 'sensitivity' model in contrast to the above 'exposure' model of tolerance. This suggests that vulnerability to nicotine dependence is related to high initial individual sensitivity to nicotine. They argue that initial exposure to nicotine in individuals who have high innate sensitivity produces more intensive reinforcing effects, and that nicotine exposure for these people quickly leads to the development of tolerance and dependence. Individuals less sensitive to nicotine initially may experience less intense effects from nicotine and will not continue to smoke because they are less responsive to its reinforcing effects.

The validity of the stage models has not yet really been established. For some girls in the London study,[3] daily smoking seemed to develop much more quickly than previously envisaged. Approximately half of those who took up smoking during the study were smoking on a daily basis within one year. It is possible that the two models may both be appropriate, and that different sensitivities to nicotine may produce different developmental patterns and determine the rate at which a teenage smoker moves through the stages. Further research is needed in this area.

5.2 Persistent and compulsive smoking in the face of smoking related ill-health

As discussed in chapters 3 and 4, the majority of adult smokers state that they would like to stop smoking, and indeed many have already tried unsuccessfully to quit. Some may take the view that such intentions to quit can be half-hearted in people who may not yet have any serious smoking-related illness or perceive themselves as having any immediate risk. This view would suggest that, if we really want to see how addictive smoking is, we should examine the smoking behaviour

of smokers diagnosed as having a condition known to be caused or
adversely affected by smoking, and who are aware that continued
smoking will worsen that condition. This section will summarise the
evidence on smoking cessation in such circumstances.

Cardiovascular disease

Smoking substantially increases the risk of cardiovascular disease. For
example, smokers are almost twice as likely as never-smokers to suffer
fatal ischaemic heart disease and four times as likely to suffer a fatal
aortic aneurysm.[25] Among smokers with an existing cardiovascular illness, smoking cessation slows disease progression and reduces the risk
of recurrence. For example, smoking cessation reduces the risk of
recurrence or premature death by about 50% in people with diagnosed
coronary heart disease.[26]

The evidence suggests that, despite widespread awareness of the
effects of smoking on their disease, and sometimes despite the provision of well designed smoking cessation interventions by health
professionals for these people, the majority of smokers who develop
cardiovascular disease are still smoking a year later. For example, in a
large trial of patients undergoing coronary artery bypass surgery,
only about a third of smokers had succeeded in stopping smoking at
10 years after medical or surgical intervention.[27] One recent UK
study[28] assessed smoking cessation one year after patients were newly
diagnosed with myocardial infarction (MI) or angina. Only 20% of
169 patients smoking in the two weeks prior to diagnosis had managed to quit smoking a year later, despite support by trained nurses
and an average of more than two quit attempts by these patients
throughout the year. Numerous studies have examined the effects of
smoking cessation interventions in cardiac patients. Although some
have found some beneficial effects of such interventions (when
compared to 'usual care' controls), the long-term success rates are
invariably well short of 50%.[29,30]

Certain forms of peripheral vascular disease are particularly closely
associated with tobacco smoking. Buerger's disease is a progressive
inflammatory occlusive disease almost exclusive to smokers, which
commonly requires surgical intervention including limb amputation.
Prognosis for Buerger's disease is considerably improved by stopping
smoking.[26] One recent study followed a cohort of 69 such patients
over the first 10 years after diagnosis.[31] All but one of the patients were
smokers at the time of diagnosis, and 84% continued to smoke thereafter. Among those who continued to smoke, 65% subsequently

required amputation, almost twice the percentage of those who stopped smoking.

Thus, although the onset of cardiovascular problems such as MI or angina initiates an increase in motivation and attempts to stop smoking, it is clear that the immediacy of the health risks is still not sufficient to enable most smokers to succeed in giving up.

Cancer

Tobacco smoking is associated with an increased risk of both respiratory and non-respiratory cancers. For example, lung cancer is approximately 20 times more likely in men who smoke than in never-smoking men, whilst pancreatic cancer is twice as likely in smokers than in never-smokers.[25] Smoking cessation reduces the risk of developing cancer, the severity and progression of premalignant histological changes in the organs such as the lung and the cervix[26] and the risk of further neoplasms, and it improves survival rates in patients who have cancer.[26,32]

Studies of smoking cessation following cancer diagnosis face the problem that some patients who believe their illness to be terminal may perceive no advantage to stopping smoking. Others may have been prevented from smoking by the severity or circumstances of their illness both before and after diagnosis. There are also problems related to underreporting of smoking by patients. Studies of smoking cessation in patients with lung cancer typically find that about 40% have succeeded in stopping smoking by two years after diagnosis and lung resection.[33] Some studies using specially designed smoking cessation interventions have found one-year abstinence rates as high as 60% in head and neck cancer patients,[34] but others have found that only 21% remain abstinent six weeks after the intervention.[35] Although the diagnosis of a smoking related malignancy clearly represents a strong catalyst for smoking cessation, it is however a consistent finding that a sizeable subgroup of patients (generally more than one-third) continue to smoke despite attempting to quit.[36]

Respiratory disease

Tobacco smoking is a major cause of a large number of respiratory illnesses, including chronic obstructive pulmonary disease (COPD).[25] The impact of smoking cessation on the progression of COPD was assessed definitively in the landmark Lung Health Study.[37] This study aimed to assess whether an intensive smoking cessation intervention and the use of a bronchodilator could slow the rate of decline in lung function in otherwise healthy smokers who already had detectable, but mild, impairment of lung function. Almost 6,000 middle-aged smokers were randomised to receive an intensive smoking cessation intervention plus a bronchodilator inhaler, the cessation intervention plus placebo inhaler, or no intervention. All participants were required to be smokers of at least 10 cigarettes per day, be willing to consider smoking cessation, and to participate in a five-year follow-up. The sample recruited had actually smoked an average of over 30 cigarettes per day for over 30 years. The smoking cessation intervention used was one of the most intensive ever evaluated in a large trial,[38] and included 'aggressive' encouragement to use (free) nicotine gum. The cessation rates achieved with this intervention were among the highest reported in a major trial, with 35% of the intervention group being abstinent at one year, compared with 9% in the non-intervention group. At five years, 22% of the intervention group were still not smoking, compared with 5% of the control group.

The high cessation rates found in this study are partly attributable to the careful screening of participants, which ensured that the study population was highly motivated to give up (as shown by the high cessation rate in the control group). However, even with this intensive intervention in motivated patients, the fact remains that more than 75% of smokers with established early COPD were still smoking five years later. Studies involving unselected patients with respiratory disease in Britain have tended to find one-year smoking cessation rates of under 10%,[39-41] and that cessation rates in patients with respiratory disease tend to be lower than in those with cardiac disease.[40]

Smoking cessation in hospital patients

The relatively low cessation rates (<10%) in the British studies described above[39-41] suggest, contrary to what might be expected, that patients with smoking related diseases are no better than other patients or non-patients at stopping smoking. Some studies have actually found that patients with smoking related diseases are less successful at stopping smoking than other types of hospital patients.[42] Consistent with this, patients in the Lung Health Study with more respiratory symptoms at baseline were less likely to succeed in stopping smoking.[43]

Recent trials of specially designed interventions for hospitalised patients have achieved one-year abstinence rates as high as 27%,[44] although it is noticeable that in such studies the one-year biochemically confirmed abstinence rates in the control (usual care) groups were often below 10%.[45]

Pregnancy

The other condition in which smokers have a clear and immediate reason for stopping smoking is pregnancy because smoking in pregnancy damages the fetus (see Section 1.4). Despite widespread awareness of the adverse health effects of smoking during pregnancy, only one in five women in the UK who are smoking when they discover they are pregnant manage to stop smoking during their pregnancy.[46] Even when specially designed smoking cessation interventions are provided, this leads at best to only an additional 10% of women stopping smoking while pregnant.[47]

Conclusions

The evidence reviewed in this section demonstrates that the onset of certain conditions such as lung cancer, a heart attack or pregnancy provides the motivation for a higher proportion of smokers to try to succeed in stopping smoking than would otherwise have been the case. However, even after developing a serious smoking related illness which threatens loss of life or limbs in the immediate future, the majority of smokers are unable to stop smoking completely in the year after their condition is diagnosed. This remains true when medical advice and nursing support are provided, and when the smokers make determined attempts to stop smoking. When considered in conjunction with the other available evidence on the role of nicotine in tobacco smoking, this provides convincing evidence that many smokers suffer serious ill-health not through personal choice, but because they are, and remain, dependent on the nicotine they obtain from tobacco.

5.3 Non-dependent smokers

In 1976, Zinberg and Jacobsen[48] used the term 'chippers' to refer to opiate users who were capable of controlling and limiting their use of opiates, as opposed to the common pattern of escalating and compulsive opiate use which many had come to associate with heroin users. Their paper is one of many in the drug addiction field which recognises that drugs which have strong dependence-producing qualities in many people do not necessarily produce dependence in all users.[49]

Since those early studies of non-dependent heroin users, it has been recognised that not all tobacco smokers progress to become highly dependent chain smokers. Shiffman, in the US, was one of the first to study systematically the phenomenon of non-dependent smokers, and he also used the term 'chippers'.[50] This section will summarise what is known about non-dependent smokers.

Definition and prevalence of non-dependent smoking

The question of what proportion of smokers is dependent is not as simple as it might at first seem to be. Degree of dependence is best conceptualised as existing on a continuum, rather than as a dichotomous variable (dependent vs non-dependent). Central to the definition of dependence is the perception of some compulsion to smoke and a sense that it would be difficult to abstain. There is good evidence that the degree of dependence (when defined as difficulty in abstaining) is closely related to the frequency of smoking.[51] In the UK, almost 58% of current smokers state that they would find it very or fairly difficult to go without smoking for only one day. People smoking 20 or more cigarettes per day are more likely to say it would be difficult than those smoking less than 10 a day (83% vs 23%).[52] The proportion of non-dependent smokers is therefore related to the definitions both of 'a smoker' and of 'dependence'.

There appears to be a consensus in the literature that adults who consistently smoke five or fewer cigarettes per day (but who smoke on at least four days per week) over a long period (eg more than a year) are non-dependent. Although cross-sectional surveys find that up to 20% of UK smokers report smoking fewer than five cigarettes per day,[52] it is unclear what proportion of these people are in a transitional phase of increasing or decreasing consumption. One study in Australia found that only 8% of 700 adult smokers smoked five or fewer cigarettes per day.[53] However, most of these were preparing to quit smoking, and so may have been reducing their consumption. It has been estimated that only about 5% of smokers are able to smoke without becoming addicted.[54]

The first studies of tobacco chippers compared them to samples of heavy smokers (20–40 cigarettes per day). The light smokers reported no signs of nicotine withdrawal after overnight abstinence and, in contrast to heavy smokers, also reported that they could regularly and easily abstain from tobacco for periods of a few days or longer.[50] This confirms that they were at the low end of the dependence continuum. However, it was also found that the chippers' nicotine absorption per cigarette and nicotine elimination rates were similar to those of heavy smokers.[55,56] The chippers were less likely both to smoke to relieve stress and to report an aversive response to their first ever cigarette. The light smokers also reported having fewer smoking relatives.

A UK study compared very light smokers (consistently less than six cigarettes per day) with regular smokers in a cohort of women followed up for a year.[57] This study found that very light smokers had higher educational attainment, more non-smoking relatives, but also more very light smokers among their relatives, lower neuroticism scores, and were less likely to state that they smoked to help them cope.

It has been suggested that vulnerability to nicotine dependence is related to genetically based high initial sensitivity to nicotine.[23] Consistent with this, people who become highly dependent cigarette smokers have been found to have more pleasurable sensations at their initial exposure to tobacco.[58] It has also been reported that regular smokers recalled more unpleasant reactions to their first cigarette than chippers.[54] Another study found that initial dizziness predicted increased likelihood of rapidly progressing to further smoking in children.[15]

Conclusions

A small proportion of smokers (probably about 5%) are able to maintain regular, but low levels of tobacco consumption and periods of abstinence with little difficulty. It is not clear why this is the case for some but not for the majority, but some available evidence is consistent with a constitutionally reduced sensitivity to nicotine. The existence of tobacco chippers does not imply that nicotine is non-addictive, any more than the existence of opiate chippers implies that heroin is non-addictive.

5.4 Smoking cessation rates in Britain

Cigarette smoking is a chronically relapsing behaviour. It is clear that many who have sustained long periods of abstinence go back to smoking. From an increasing group of cessation trials with long-term follow-up we know that relapse rates among those initially succeeding decline over time, but still remain substantial for some years: about 50% of those abstinent for six weeks relapse by six months, and 20% of those abstinent for six months relapse by one year.[59] There is also evidence that 30–50% of those abstinent for one year will relapse before five years.[60,61]

It might be tempting to think that such trial data give a pessimistic view of relapse because their populations include more dependent smokers, frequently with a long history of failure. However, at least one major population survey recording retrospective reports of periods of abstinence supports these results, giving a figure of 35% relapse after one year of abstinence.[26] The same survey suggests that an asymptote in relapse rates may be reached after about five years of cessation – but until an ex-smoker dies there can be no certainty that he or she has 'quit for life', which makes it extremely difficult accurately to calculate lifetime cessation rates at any one point in time.

The 'annual cessation rate' is the percentage of regular smokers in a population who quit each year and remain abstinent. It can be aggregated over a long period to estimate the 'lifetime cessation rate': that is, the percentage of ever-smokers who stop and stay stopped. These statistics, together with the rate of smoking uptake among never-smokers, broadly define the dynamics of smoking prevalence. Unfortunately, there are few prospective studies of the natural history of smoking within individuals to allow these rates to be estimated reliably in the British population, but existing cross-sectional data collected repeatedly over time in the General Household Survey (GHS)[52] can be used to provide estimates of cessation rates.

Figure 5.1 shows the cross-sectional smoking prevalence curves plotted by age for the 13 surveys containing smoking data between 1972 and 1996. There are two striking features in these data:

1   The curves demonstrate a consistent trend for smoking prevalence to fall within the older age groups. The only major exception is the curve for 1996, which shows stable or increasing prevalence at all ages.

2   There is no evidence of a similar trend towards reduced prevalence over time in those under the age of 25, since the prevalence curves for this age group are tightly clustered from 1982 onwards. As discussed in Section 1.2, this reflects the fact that smoking prevalence

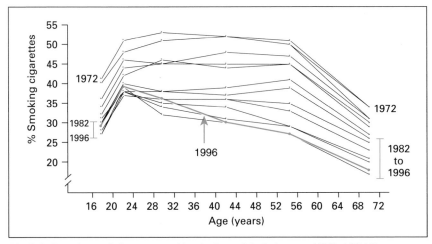

Fig 5.1. *Prevalence of cigarette smoking in Great Britain by age, 1972–1996.*[52]

in younger people has been stable or increasing in recent years. Very few people start smoking after the age of 25, so it is reasonable to infer that until the early 1980s the declining prevalence in Britain was due, at least in part, to falling rates of uptake of smoking, but that the decline since then is attributable primarily to the loss of prevalent smokers through smoking cessation or death.

Estimation of annual cessation rates

There are several options for estimating annual cessation rates over time from the GHS cross-sectional surveys, although none is ideal. One method is to calculate cessation rates directly as the percentage of those who have ever smoked who now regard themselves as ex-smokers. This ratio has been used extensively in the past, but may be subject to inaccuracies in respondents' classification of themselves as ex-smokers.[62]

An alternative approach is to follow the smoking prevalence among members of a particular age cohort as they are sampled by the GHS over time. Few people start smoking beyond the age of 25, few regular smokers quit before the age of 25, and the excess mortality from smoking does not have its most substantial impact until after the age of 60, so relatively reliable estimates of cessation rates can be obtained for those in the age range 25–59 years. Smoking cessation in this population is particularly important from a public health perspective, because quitting during this time enables a large proportion of the life-years to be saved that would otherwise be lost by continued smoking.[63]

Those aged 25–49 in the 1986 survey were sampled again by the GHS at the age of 35–59 in the most recent survey in 1996–1997. Figure 5.2 shows their estimated cigarette smoking prevalence at the two time points. Prevalence fell from about 36% to 28.8% over the 10-year period, indicating a long-term net reduction in smoking, after allowing for relapse. Assuming that the cessation rate remained constant over this time, the 7.2% (95% confidence intervals (CI) 5.7–8.7) reduction equals an annual cessation rate of about 2.2% (95% CI 1.7–2.8). The rate was similar for men (2.4%) and women (2.0%). It was lower for the younger group, aged 25–34 in 1986 (ca 1.4% per year), and higher for the older group, aged 35–49 in 1986 (ca 2.6%). Over the 10-year period, 20% of those aged 25–49 in 1986 gave up smoking cigarettes long-term, possibly for life.

A similar approach can be used to estimate cessation during the previous 10 years, 1976–1986. In 1976, the smoking prevalence among those aged 25–49 was 46.8%; by the time they were aged 35–59 in

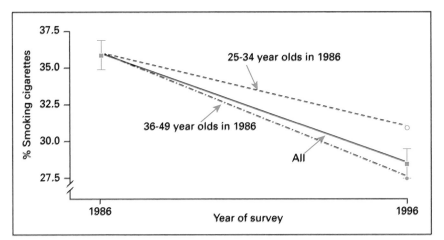

Fig 5.2. Prevalence of cigarette smoking in those aged 25–49 in 1986 and 35–59 in 1996.[63] Error bars give approximate 95% confidence intervals.

1986, it was 35.6%. The 11% reduction equals an annual cessation rate of about 2.7%, suggesting that cessation rates were higher during this earlier period than in the last 10 years. Between 1976 and 1986, 24% of smokers in this age cohort gave up.

These estimates do not take into account any smokers who switched from smoking cigarettes to cigars, and who will be misclassified as ex-smokers in this analysis. Switching from cigarette to cigar smoking tends to occur primarily in men, so this effect may have contributed to the greater cessation rate observed in men relative to women, but in absolute terms the effect is likely to be very small. The prevalence of cigar smoking in people who have switched from cigarettes has also remained fairly constant in this population, so adjusting for this effect would have little influence on overall cessation estimates.

Cessation before the age of 60

Those aged 25–34 in 1972 can be followed until they were aged 50–59 in 1996–1997 to estimate total cessation before the age of 60. Cigarette smoking prevalence in this population fell from 53% to 27% over the 25 years, suggesting that 49% stopped before the age of 60. This esti-mate is possibly inaccurate because it does not allow for permanent switching to cigars or the fact that the majority were not followed until the age of 60, but it is in close agreement with previously published figures indicating that only about 50% of smokers stop smoking long-term before the age of 60.[62]

Conclusions

In the last 10 years, long-term cessation rates among middle-aged smokers have averaged around 2% per year, ranging from about 1.4% to 2.6% with increasing age. This is somewhat less than the 2.7% per year estimated for the previous 10 years, suggesting that cessation has been slowing. Over the last 25 years, about 50% of smokers have stopped smoking before the age of 60.

Overall, the evidence in this chapter demonstrates that in most smokers, dependence on nicotine is established during adolescence, and in many cases probably occurs before reaching the legal minimum age for the purchase of cigarettes. Once dependence is established, the majority of smokers will then continue to smoke for nearly 40 years. Experimentation with cigarettes in adolescence clearly has major long-term implications for individual and public health.

References

1 Prokhorov AV, Pallonen UE, Fava JL, Ding L, Niaura R. Measuring nicotine dependence among high-risk adolescent smokers. *Addict Behav* 1996; **21**: 117–27.

2 Stanton RW. DSM-III-R Tobacco dependence and quitting during late adolescence. *Addict Behav* 1995; **20**: 595–603.

3 McNeill AD, Jarvis MJ, Stapleton JA, West RJ, Bryant A. Nicotine intake in young smokers: longitudinal study of saliva cotinine concentrations. *Am J Pub Health* 1989; **79**: 172–5.

4 Goddard E. *Why children start smoking*. London: HMSO, 1990.

5 Hedges B, Jarvis MJ. Cigarette smoking. In: Prescott-Clarke P, Primatesta P (eds). *Health survey for England. The health of young people '95–'97*. London: The Stationery Office, 1998: 191–222.

6 McNeill AD, Jarvis MJ, West R, Russell MAH, Bryant A. Saliva cotinine as an indicator of cigarette smoking in adolescents. *Br J Addict* 1987; **82**: 1355–60.

7 McNeill AD, West R, Jarvis M, Jackson P, Bryant A. Cigarette withdrawal symptoms in adolescent smokers, *Psychopharmacology* 1986; **90**: 533–6.

8 Barton J. *Young teenagers and smoking in 1997. A report of the key findings from the Teenage Smoking Attitudes Survey carried out in England in 1997*. London: Office for National Statistics, 1998.

9 Lynch BS, Bonnie RJ. *Growing up tobacco free: preventing nicotine addiction in children and youths*. Institute of Medicine. Washington, DC: National Academy Press, 1994.

10 Barker D. Reasons for tobacco use and symptoms of nicotine withdrawal among adolescent and young adult tobacco users – United States, 1993. *MMWR* 1994; **43**: 745–50.

11 Sargent JD, Mott LA, Stevens M. Predictors of smoking cessation in adolescents. *Arch Pediatr Adolesc Med* 1998; **152**: 388–93.

12 McNeill AD, Jarvis M, West R. Subjective effects of cigarette smoking in adolescents. *Psychopharmacology* 1987; **92**: 115–7.

13 McNeill AD. The development of dependence on smoking in children. *Br J Addict* 1991; **86**: 589–92.

14 Benowitz NL. Pharmacologic aspects of cigarette smoking and nicotine addiction. *N Engl J Med* 1988; **319**: 1318–30.

15 Hirschman RS, Leventhal H, Glynn K. The development of smoking behaviour: conceptualisation and supportive cross-sectional survey data. *J Appl Soc Psychol* 1984; **14**: 184–206.

16 Friedman LS, Lichtenstein E, Biglan A. Smoking onset among teens: an empirical analysis of initial situations. *Addict Behav* 1985; **10**: 1–13.

17 Glynn K, Leventhal H, Hirschman R. A cognitive developmental approach to smoking prevention. In: Bell CS, Battjes R (eds). *Prevention research: deterring drug abuse among children and adolescents.* NIDA Research Monograph 63. Washington, DC: US Government Printing Office, 1985.

18 McNeill AD, Jarvis MJ, Stapleton JA, Russell MAH, *et al.* Prospective study of factors predicting uptake of smoking in adolescents. *J Epidemiol Community Health* 1989; **43**: 72–8.

19 Russell MAH. Cigarette smoking: natural history of a dependence disorder. *Br J Med Psychol* 1971; **44**: 1–16.

20 McKennell AC, Thomas RK. *Adults and adolescents' smoking habits and attitudes.* London: HMSO, 1967.

21 Flay BR, d'Avernas JR, Best JA, Kersell MW, Ryan KB. Cigarette smoking: why young people do it and ways of preventing it. In: McGrath PJ, Firestone P (eds). *Pediatric and adolescent behavioural medicine.* New York: Springer-Verlag, 1983: 132–83.

22 Russell MAH. The smoking habit and its classification. *Practitioner* 1974; **212**: 791–800.

23 Pomerleau OF, Collins AC, Shiffman S, Pomerleau CS. Why some people smoke and others do not: new perspectives. *J Consult Clin Psychol* 1993; **61**: 723–31.

24 Pomerleau OF. Nicotine dependence. In: Bolliger CT, Fagerström KO (eds). *The tobacco epidemic. Progress in Respiratory Research* 1997; **28**: 122–31. Basel: Karger.

25 Wald NJ, Hackshaw AK. Cigarette smoking: an epidemiological overview. *Br Med Bull* 1996; **52**: 3–11.

26 US Department of Health and Human Services. *The health benefits of smoking cessation: a report of the Surgeon-General.* US DHHS Publication No. (CDC) 90-8416. Rockville, MD: USDHHS, 1990.

27 Rogers WJ, Coggin CJ, Gersh BJ, Fisher LD, *et al.* Ten year follow-up of quality of life in patients randomized to receive medical therapy or coronary bypass graft surgery. The Coronary Artery Surgery Study (CASS). *Circulation* 1990; **82**: 1647–58.

28 Jolly K, Bradley F, Sharp S, Smith H, *et al.* Randomised controlled trial of follow up care in general practice of patients with myocardial infarction and angina: final results of the Southampton Heart Integrated Care Project (SHIP). *Br Med J* 1999; **318**: 706–11.

29 Johnson JL, Budz B, Mackay M, Miller C. Evaluation of a nurse-delivered smoking cessation intervention for hospitalised patients with cardiac disease. *Heart Lung* 1999; **28**: 55–64.

30 Cupples ME, McNight A. Randomised controlled trial of health promotion in general practice for patients at high cardiovascular risk. *Br Med J* 1994; **309**: 993–6.

31 Borner C, Heidrich H. Long-term follow-up of thromboangiitis obliterans. *Review Vasa* 1998; **27**: 80–6.

32 Gritz ER. Smoking and smoking cessation in cancer patients. *Br J Addict* 1991; **86**: 549–54.

33 Gritz ER, Nisenmaum R, Elashoff RE, Holmes EC. Smoking behavior following diagnosis in patients with stage 1 non-small cell lung cancer. *Cancer Causes Control* 1991; **2**: 105–12.

34 Gritz ER, Carr CR, Rapkin D, Abemayor E, *et al.* Predictors of long-term smoking cessation in head and neck cancer patients. *Cancer Epidemiol Biomarkers Prev* 1993; **2**: 261–70.

35 Griebel B, Wewers ME, Baker CA. The effectiveness of a nurse-managed minimal smoking-cessation intervention among hospitalised patients with cancer. *Oncol Nurs Forum* 1998; **25**: 897–902.

36 Ostroff JS, Jacobsen PB, Moadel AB, Spiro RH, *et al.* Prevalence and predictors of continued tobacco use after treatment of patients with head and neck cancer. *Cancer* 1995; **75**: 569–76.

37 Anthonisen NR, Connett JE, Kiley JP, Altose MD, *et al.* Effects of smoking intervention and use of an inhaled anticholinergic bronchodilator in the rate of decline of FEV1. The Lung Health Study. *JAMA* 1994; **272**: 1497–505.

38 Foulds J. Strategies for smoking cessation. *Br Med Bull* 1996; **52**: 157–73.

39 British Thoracic Society. Comparison of four methods of smoking withdrawal in patients with smoking related diseases. *Br Med J* 1983; **286**: 595–7.

40 British Thoracic Society. Smoking withdrawal in hospital patients: factors associated with outcome. *Thorax* 1984; **39**: 651–6.

41 British Thoracic Society. Smoking cessation in patients: two further studies by the British Thoracic Society. *Thorax* 1990; **45**: 835–40.

42 Foulds J, Stapleton J, Hayward M, Russell MAH, *et al.* Transdermal nicotine patches with low-intensity support to aid smoking cessation in outpatients in a general hospital: a placebo-controlled trial. *Arch Fam Med* 1993; **2**: 417–23.

43 Nides MA, Rakos RF, Gonzales D, Murray RP, *et al.* Predictors of initial cessation and relapse through the first two years of the Lung Health Study. *J Consult Clin Psychol* 1995; **63**: 60–9.

44 Miller NH, Smith PM, DeBusk RF, Sobel DS, Taylor CB. Smoking cessation in hospitalized patients. Results of a randomized trial. *Arch Intern Med* 1997; **157**: 409–15.

45 Simon JA, Solkowitz SN, Carnody TP, Browner WS. Smoking cessation after surgery. A randomized trial. *Arch Intern Med* 1997; **157**: 1371–6.

46 Owen L, McNeill A, Callum C. Trends in smoking during pregnancy in England, 1992–7: quota sampling surveys. *Br Med J* 1998; **317**: 728–30.

47 West R. Interventions to reduce smoking among pregnant women. In: *Health Education Authority: smoking and pregnancy. Guidance for purchasers and providers.* London: HEA, 1994.

48 Zinberg NE, Jacobsen RC. The natural history of 'chipping'. *Am J Psychiatry* 1976; **133**: 37–40.

49 Powell DG. A pilot study of occasional heroin users. *Arch Gen Psychiatry* 1973; **28**: 586–94.

50 Shiffman S. Tobacco 'chippers – individual differences in tobacco dependence. *Psychopharmacology* 1989; **97**: 539–47.

51   Etter JF, Duc TV, Perneger TV. Validity of the Fagerström test for nicotine dependence and of the Heaviness of Smoking Index among relatively light smokers. *Addiction* 1999; **94**: 269–81.

52   Thomas M, Walker A, Wilmot A, Bennett N. *Living in Britain: results from the 1996 General Household Survey.* Office for National Statistics. London: The Stationery Office, 1998.

53   Owen N, Kent P, Wakefield M, Roberts L. Low-rate smokers. *Prev Med* 1995; **24**: 80–4.

54   Shiffman S. Refining models of dependence: variations across persons and situations. *Br J Addict* 1991; **86**: 611–5.

55   Shiffman S, Fischer LB, Zettler-Segal M, Benowitz NL. Nicotine exposure among nondependent smokers. *Arch Gen Psychiatry* 1990; **47**: 333–6.

56   Shiffman S, Zettler-Segal M, Kassel J, Paty J, *et al.* Nicotine elimination and tolerance in non-dependent cigarette smokers. *Psychopharmacology* 1992; **109**: 449–56.

57   Hajek P, West R, Wilson J. Regular smokers, lifetime very light smokers, and reduced smokers: comparison of psychosocial and smoking characteristics in women. *Health Psychol* 1995; **14**: 195–201.

58   Pomerleau OF, Pomerleau CS, Namenek RJ. Early experiences with tobacco among women smokers, ex-smokers, and never-smokers. *Addiction* 1998; **93**: 595–9.

59   Stapleton JA. Cigarette smoking prevalence, cessation and relapse. *Stat Methods Med Res* 1998; **7**: 187–203.

60   Stapleton JA, Sutherland G, Russell MAH. How much does relapse after one year erode effectiveness of smoking cessation treatments? Long term follow up of randomized trial of nicotine nasal spray. *Br Med J* 1998; **316**: 830–1.

61   Blondal T, Gudmundsson LJ, Olafsdottir I, Gustavsson G, Westin A. Nicotine nasal spray with nicotine patch for smoking cessation: randomized trial with six year follow up. *Br Med J* 1999; **318**: 285–8.

62   Jarvis MJ. Patterns and predictors of smoking cessation in the general population. In: Bolliger CT, Fagerström KO (eds). *The tobacco epidemic. Progress in Respiratory Research* 1997; **28**: 151–64. Basel: Karger.

63   Doll R, Peto R, Wheatley K, Gray R, Sutherland I. Mortality in relation to smoking: 40 years' observations on male British doctors. *Br Med J* 1994; **309**: 901–11.

6 | Regulation of nicotine intake by smokers, and implications for health

6.1 Titration and compensation
6.2 Epidemiological evidence on the effects of changes in tar and nicotine yields of cigarettes on disease risk
6.3 Does the epidemiological evidence support a direct relation between reduction in cigarette yields and disease risk?

6.1 Titration and compensation

The extent to which tobacco users regulate their nicotine intake has important implications both for our understanding of the factors driving smoking behaviour and for attempts to design less toxic smoking products acceptable to smokers. A number of methodologies have been used to examine this issue, including:

- study of the effects of intravenous nicotine, nicotine chewing gum and nicotine patches on ongoing smoking behaviour[1]
- comparison of nicotine intakes among users of different kinds of tobacco products, combustible and non-combustible, which differ markedly in sensory aspects but have in common only nicotine delivery,[2,3] and
- comparison of nicotine intakes from smoking cigarettes with low, medium and high nicotine deliveries.[1,4]

This research literature has provided abundant evidence for the underlying role of nicotine in smoking behaviour. Because the delivery of both tar and nicotine is highly correlated in cigarette smoke,[5] there has been concern about the difficulty of distinguishing effects of tar and of sensory factors from effects of nicotine on smoker compensation. This concern has been largely met by the consistency of findings from observational and experimental studies and from differing tobacco products and routes of delivery. The failure in the market-place of nicotine-free cigarettes delivering substantial levels of tar has also provided powerful indirect evidence.

It is now well established that users of tobacco tend to regulate or titrate their nicotine intake to maintain body levels within a certain range, but there are a number of unanswered questions about the determinants and stability of this range of levels both within and between individuals:

1　What determines the wide variation in nicotine intake observed between individuals?

2　To what extent are preferred levels of nicotine intake set by needs intrinsic to the smoker or by the level of nicotine delivery available from the cigarettes smoked in the early stages of uptake of the habit? For example, will smokers who took up smoking in the 1950s, when nominal nicotine deliveries averaged over 2 mg per cigarette, have had their preferred intakes set at a higher level than 1990 novice smokers exposed to cigarettes delivering 1 mg or less?

Definitive answers to most of these questions are still largely lacking. Of particular significance from a public health perspective is the issue of smoker compensation from cigarettes with differing nominal nicotine yields. The process whereby smokers change the way they smoke lower yield cigarettes to compensate for their lower output raises further important questions:

1　At a given point in time, what is the evidence that, within individual smokers, those smoking brands with higher or lower machine-smoked nicotine yields take in different nicotine doses?

2　If smokers switch to a cigarette brand with a lower nicotine delivery, to what extent do they increase their intake of nicotine per cigarette to compensate for the reduction in delivery?

3　What is the evidence that average nicotine intakes in the population have declined over time as cigarette nicotine deliveries have been progressively reduced?

Such questions have been addressed by several types of research study. Naturalistic studies of own brand smoking have examined biochemical measures of smoke intake in groups who have self-selected to smoke brands with differing nominal deliveries. Experimental studies of brand switching (so called forced-switching studies) have looked at the consequences, usually short-term, of changing from usual brand to a lower yielding cigarette (or, less frequently, to a higher yielding brand). In a very few studies, the natural history of brand switching has been examined: that is, subjects have been followed over time, and biochemical measures of intake made before and after spontaneous brand changes. Each type of research design presents particular problems of interpretation, but taken together they provide a consistent picture.

Cross-sectional studies of self-selected own brand smoking

Since the early 1980s, a number of studies have examined smoke intake from own brand smoking, mostly using measures of saliva or plasma cotinine as an indicator of total nicotine intake.[6–16] Some studies have looked at particular groups of smokers, such as those attending smokers' clinics,[6] while others have been population based.[11,16] Studies have also varied in the proportion of subjects smoking ultra-low yielding cigarettes (<0.5 mg nicotine), which remain a rare choice in the general population of smokers. A consistent finding has been that, while nicotine intakes vary greatly between individuals smoking brands of any given delivery, there is relatively little difference in average nicotine intake from brands with differing yields. Some studies have found no differences across the range of yields examined, whilst others have observed a slight, but significant trend to lower intake with lower yields. In yet others, intakes have been flat across a wide range of yields and reduced only in subjects smoking ultra-low brands.

These studies are valuable because they demonstrate a reliable tendency for smokers to over-smoke brands with low deliveries and under-smoke those with high nicotine yields, making nominal machine-smoked yields of little or no value in predicting smokers' actual intakes. They establish that a high degree of nicotine compensation is certainly the norm, and indeed do not exclude the possibility that complete compensation is the rule. The observation of lower intakes in ultra-low yield smokers is more consistent with an effect of reduced yield, but this effect could be due to self-selection of lower yielding brands in lighter smokers with lower preferred intakes who do not find the degree of compensatory increased puffing required too aversive.

Forced-switching studies

Most experimental studies of brand switching have been short-term, lasting only a few days or weeks.[4] Typically, smokers' intakes of nicotine and other smoke components have been substantially reduced after switching to a lower yielding brand, although not to the extent predicted by the degree of yield reduction.[1,17–19] A problem with this type of study is that, although subjects compensate only partially, they also find the low yielding cigarettes unsatisfying. This makes it hazardous to generalise to the real-world situation. It is entirely plausible that, outside the experimental situation, smokers would not persist with cigarettes they found unsatisfying. Data from forced-switching studies, therefore, provide a lower bound for the extent of smoker compensation, but not an upper bound.

Two large switching studies in the UK are unique in that they have randomised large numbers of smokers to lower yielding brands and followed them over a period of six months.[20,21] In one study,[20] nicotine compensation was essentially complete, while in the other[21] it was estimated to be about 80%.

Spontaneous brand switching

Two published studies have followed smokers over time and measured changes in intake of nicotine and other smoke components in relation to spontaneous brand switching. In one study,[22] comparable numbers of smokers switched from higher to lower yielding brands, and from lower to higher. Those who switched down had previously been smoking cigarettes with relatively high nicotine yields, while those who switched up had initially smoked low yielding cigarettes (0.42 mg nicotine). In each case measured, intakes of nicotine and carbon monoxide changed in the same direction as the change in nominal yield from the cigarette. However, the reduction in intake observed in smokers who switched to a lower yielding brand was achieved not by a reduction in nicotine intake per cigarette – which in fact remained relatively constant – but by a reduction in the number of cigarettes smoked. The implication, therefore, is that the change to a brand with a lower yield may have been driven by a desire to cut down on smoking in general, in which a change to a lower yield brand was perceived to be helpful but in reality did not contribute to the reduction in intake.

The second study[23] investigated the effect of change in brand yield on rate of decline in lung function (measured as FEV_1) in men followed over a period of 13 years. Concentrations of nicotine metabolites in the urine were similar over time in all the subjects, whether or not they had switched to a lower tar group, indicating complete compensation of intake.

The overall implication of these studies is therefore that smokers who switch to lower yield brands of cigarette tend to compensate for this change by smoking their cigarettes in a way that provides a greater intake of nicotine. They thus succeed in titrating their body levels to their own internally defined desired range. If this is the case, the question arises whether any realistic benefit in health terms can be achieved, or could reasonably be expected, by encouraging smokers to switch to lower yielding brands.

6.2 Epidemiological evidence on the effects of changes in tar and nicotine yields of cigarettes on disease risk

The manufactured cigarette of today is a very different product from that on sale 40 years ago. The filter tip became widespread in the 1950s, and standard length cigarettes have been successively replaced by king size and super kings. Changes in tobacco processing technology have included the widespread use of both reconstituted tobacco sheet and expanded tobacco, as well as additives of various kinds. Machine-smoked estimates of tar and nicotine yields, using methods such as the US Federal Trade Commission (FTC) protocol (see Section 8.2), have declined continuously and substantially from over 30 mg tar and 2.5 mg nicotine to about 11 mg tar and 1 mg nicotine. In the UK, these continuing yield reductions were the subject of a series of voluntary agreements between the tobacco industry and government, now superseded by regulation at the European level.[24] It is important to note, however, that the yield reductions have not been due primarily to changes in the underlying tobacco content (indeed, the nicotine content of the tobacco in ultra-low yielding brands may actually be higher than that of high yielding brands), but have been achieved principally by the introduction of filter ventilation holes to dilute the smoke inhaled from the cigarette with a variable amount of air.[25] This results in a reduction in machine-smoked tar and nicotine yields measured by the FTC or similar protocols, but can be readily defeated by a smoker seeking higher intakes. As discussed above, there is overwhelming evidence that, in practice, smokers achieve higher intakes from lower yielding brands than those predicted from machine smoking.

Lowering tar and nicotine yields has been the main plank of product modification policies aimed at reducing the health risks of cigarettes. The question that arises, particularly in view of compensatory smoking for reduced nicotine, is whether lowered yields have produced a net benefit to public health. The answer must take into account not only any reduction of smoking related disease resulting from lowered yields, but also any possible adverse effects attributable to the adoption of low yield brands as opposed to giving up smoking completely.[26,27]

The relation between cigarette yield and lung cancer risk

A number of epidemiological studies, both case-control and prospective, have examined the relationship of the tar and nicotine yield of the brand of cigarette smoked with the risk of lung cancer. In some studies, smokers of filter cigarettes have been compared with those smoking regular untipped brands,[28-31] while in others an attempt has been made to categorise smokers by the specific tar yield of their brands.[32,33] Findings have consistently pointed to a lowering of risk in smokers of filter by comparison with untipped cigarettes, and with lower compared with higher tar yield. The relative risk in smokers of non-filter compared to filter cigarettes has ranged from 1.3 to 2.3 in different studies.[34] Tang et al[33] estimated that the relative mortality from lung cancer for a 15 mg decrease in tar yield per cigarette was 0.75. Stellman and Garfinkel[32] concluded that the excess lung cancer risk for current smokers was directly proportional to the estimated total milligrams of tar consumed daily (standardised mortality ratio 100 + 1.731 × mg of tar per day). Reviewing this literature in 1986,[34] Stellman noted that most smokers at that time had almost invariably smoked untipped cigarettes before switching to lower yielding filter brands, and argued that in future cohorts of smokers, who would be exposed to cigarettes of much lower yield for much greater proportions of their lives, the associated risks would decline even further.

An analysis which considered time trends in lung cancer death rates in British smokers over the period in which cigarette yields were declining rapidly came to a similarly optimistic conclusion.[35] Lung cancer death rates in men aged 35–44 declined by over 50% over the 25 years to 1983. Since changes in cigarette smoking prevalence could account for only a small proportion of this, it was inferred that the most likely cause of the decrease was the change in tar yields. On this basis, Peto concluded that the introduction of cigarette tar level reduction in countries where tar levels remain high might ultimately avoid about half of all cigarette-induced lung cancer.[35] More recent data from the USA, however, do not support this interpretation. Results from the American Cancer Society's (ACS) second major cohort CPS2, conducted in the 1980s, suggest that there has been no reduction in lung cancer risk relative to the earlier CPS1 baseline assessment in 1959. Rates of lung cancer in men aged under 55 were in fact somewhat higher in the more recent cohort, despite reductions in the tar yield of cigarettes in both the USA and the UK.[36] It is therefore not clear whether the trend in British lung cancer data is indeed attributable to the change in cigarette tar yields.

Cigarette yields and other cancers

The pattern of findings for cancers other than lung cancer (including oesophageal, larynx and bladder cancers) is more consistent, with studies tending to demonstrate an association between the use of filter or lower yield brands and a lower risk of disease.[34,37-39]

Heart disease. Data from the ACS prospective study which was begun in 1959 showed that deaths from coronary heart disease rose consistently with yield of cigarette smoked.[40] Parish *et al*[41] reported that rates of non-fatal myocardial infarction were slightly (10%) higher in medium tar than low tar cigarette smokers, but noted the difficulty of making adequate adjustment in their data for confounding by socio-economic factors.

Chronic obstructive pulmonary disease. A review of data from cross-sectional and prospective studies indicated that smoking low yield cigarettes leads to lower phlegm, reduced cough and less shortness of breath, but found no evidence for an effect on mortality.[40] As stated above, the rate of decline in lung function among smokers followed up for 13 years from 1971–1983 was similar in those who switched to lower yielding brands and those whose tar group did not change.[23]

6.3 Does the epidemiological evidence support a direct relation between reduction in cigarette yields and disease risk?

Despite the general epidemiological consensus, reinforced by the conclusions of an international workshop,[42] that lowered tar yields result in lowered risks of smoking related disease, there are good reasons for questioning whether the observed associations are truly causal rather than the result of uncontrolled confounding.

First amongst these concerns is the uncertainty whether smokers of cigarettes of differing tar yields at a given point in time have substantially different tar exposures, or whether, within individuals, changes over time in nominal tar yields of the brands smoked actually result in lowered tar exposure. There is also the difficulty of assigning any individual to a particular tar category on the basis of brand smoked at one point in time when that individual will have a previous history of smoking higher yielding brands, and may indeed smoke lower yielding cigarettes in the future.

The most important concern is the difficulty of controlling adequately for factors which influence both choice of cigarette brand, and consequently yield, and the risk of smoking related disease. Principal among these is socio-economic status, which is strongly associated with brand choice, with poorer smokers being much more likely to choose higher yielding brands.[43,44] There is also strong emerging evidence that poor smokers have higher nicotine intakes than affluent smokers at any level of cigarette consumption, and hence would be expected to be more at risk of smoking related disease.[45] While most available studies have adjusted for age and sex, it is acknowledged that few have been able to examine other potential biases in subject selection or to control for differences in socio-economic status between smokers of cigarettes with different yields.[34]

The difficulty of drawing firm conclusions from comparisons between self-selected groups smoking lower or higher yielding brands can be illustrated by the Bross and Gibson study.[28] These authors used a case-control design to examine lung cancer risk in men who, at assessment in the early 1960s, smoked either filter or regular untipped cigarettes, and found a relative risk of 1.7 in those smoking the latter. Filter cigarettes became widely available only in the 1950s, so it can be inferred that for the great majority of their smoking careers the filter smokers would have smoked higher yielding plain brands. Moreover, since the median date of switching to filters in the USA was 1964,[46] those who switched before this date would have been among the group of early innovators. Such a group would be expected to differ from others in many respects, including socio-economic status, education and awareness of health risks, and even level of nicotine dependence[43] – characteristics that would also relate to risk of smoking related disease. These considerations indicate that it would be premature and unwise to interpret the association of modestly lowered risk of lung cancer, other cancers or heart disease in self-selected groups smoking filter or low tar cigarettes as causally related to lowered tar yields.

Observed lowered age-specific risks of lung cancer over time in a population (after adjusting appropriately for any changes in smoking prevalence) might be regarded as providing stronger evidence for an effect of yield reduction on risk, but here again there are problems. The fact that the secular reduction in age-specific lung cancer risk observed in the UK has not been replicated in the USA,[36] where cigarette tar yields have also reduced, suggests that other factors may be responsible. Declines in disease risk over time can also be plausibly explained by other factors, such as the reduction in tar-to-nicotine ratio in smoke that has accompanied the shift towards lower yielding brands.[24,47] This implies that, even with complete nicotine compensation, smokers would be exposed to somewhat less tar in the 1980s and 1990s than in the 1950s and 1960s. Changes in risk over time could also be due to qualitative changes in the carcinogenicity of tar, if alterations in tobacco processing technology resulted in tar which on a gram-per-gram basis was less carcinogenic. Data on this question are largely lacking from recent years, although there is evidence that smoke from cigarettes made from reconstituted tobacco sheet may be less carcinogenic.[48]

Thus, whilst it is possible that changes in cigarette yields can translate into some degree of health benefit, the effects of titration and compensation by smokers mean that any reduction in health risk achieved with low yield cigarettes is unlikely to be proportionately related to the magnitude of the nicotine yield reduction. If there is a health benefit to be accrued from smoking low rather than high yield cigarettes, this is likely to arise from reductions in tar-to-nicotine ratios rather than from changes in absolute levels of either constituent alone. There is in fact no strong evidence that actual tar exposures (as opposed to tar yields) have come down at all since the 1960s. As tobacco industry scientists speculated in an internal memorandum dating from 1984:

> Consumers may have been obtaining 14–16 mg PMWNF [particulate matter water and nicotine free – a technical term for tar] (and normal equivalent nicotine delivery) for a very long time, i.e. compensating down to 16 mg when cigarettes delivered 25 mg and compensating up if they are now smoking a 13 mg … The discussion was based on example using PMWNF but it is accepted that nicotine is both the driving force and the signal [as impact] for compensation in human smoking behaviour'.[49]

While the tobacco industry has long understood the implications of smoker compensation, there is clear evidence that smokers misunderstand published machine-smoked yields and derive false health reassurance from them,[27] as may have been intended by tobacco companies.[50] There are therefore reasonable grounds for concern that low tar cigarettes offer smokers an apparently healthier option while providing little if any true benefit. In so far as low yield cigarettes may discourage smokers who would otherwise have given up smoking completely from doing so, they may indeed be counterproductive in terms of public health.[27]

References

1   Russell MAH. Nicotine and the self-regulation of smoke intake. In: Wald N, Froggatt P (eds). *Nicotine, smoking and the low tar programme.* Oxford: Oxford University Press, 1989: 151–69.

2   Holm H, Jarvis MJ, Russell MAH, Feyerabend C. Nicotine intake and dependence in Swedish snuff takers. *Psychopharmacology* 1992; **108**: 507–11.

3   Russell MAH, Jarvis MJ, Devitt G, Feyerabend C. Nicotine intake by snuff users. *Br Med J* 1981; **283**: 814–7.

4   Benowitz NL. Dosimetric studies of compensatory cigarette smoking. In: Wald N, Froggatt P (eds). *Nicotine, smoking and the low tar programme.* Oxford: Oxford University Press, 1989: 133–50.

5   Hasenfratz M, Baldinger B, Battig K. Nicotine or tar titration in cigarette smoking behavior? *Psychopharmacology* 1993; **112**: 253–8.

6   Russell MA, Jarvis M, Iyer R, Feyerabend C. Relation of nicotine yield of cigarettes to blood nicotine concentrations in smokers. *Br Med J* 1980; **280**: 972–6.

7   Russell MA, Jarvis MJ, Feyerabend C, Saloojee Y. Reduction of tar, nicotine and carbon monoxide intake in low tar smokers. *J Epidemiol Community Health* 1986; **40**: 80–5.

8   Benowitz NL, Hall SM, Herning RI, Jacob P, *et al.* Smokers of low-yield cigarettes do not consume less nicotine. *N Engl J Med* 1983; **309**: 139–42.

9   Benowitz NL, Jacob P 3d, Yu L, Talcott R, *et al.* Reduced tar, nicotine, and carbon monoxide exposure while smoking ultralow- but not low-yield cigarettes. *JAMA* 1986; **256**: 241–6.

10 Wald NJ, Boreham J, Bailey A. Relative intakes of tar, nicotine, and carbon monoxide from cigarettes of different yields. *Thorax* 1984; **39**: 361–4.

11 Woodward M, Tunstall Pedoe H. Self-titration of nicotine: evidence from the Scottish Heart Health Study. *Addiction* 1993; **88**: 821–30.

12 Woodward M, Tunstall Pedoe H. Do smokers of lower tar cigarettes consume lower amounts of smoke components? Results from the Scottish Heart Health Study. *Br J Addict* 1992; **87**: 921–8.

13 Gori GB, Lynch CJ. Analytical cigarette yields as predictors of smoke bioavailability. *Regulatory Toxicol Pharmacol* 1985; **5**: 314–26.

14 Byrd GD, Davis RA, Caldwell WS, Robinson JH, deBethizy JD. A further study of FTC yield and nicotine absorption in smokers. *Psychopharmacology* 1998; **139**: 291–9.

15 Maron DJ, Fortmann SP. Nicotine yield and measures of cigarette smoke exposure in a large population: are lower-yield cigarettes safer? *Am J Pub Health* 1987; **77**: 546–9.

16 Coultas DB, Stidley CA, Samet JM. Cigarette yields of tar and nicotine and markers of exposure to tobacco smoke. *Am Rev Respir Dis* 1993; **148**: 435–40.

17 Russell MAH, Sutton SR, Iyer R, Feyerabend C, Vesey CJ. Long-term switching to low-tar low-nicotine cigarettes. *Br J Addict* 1982; **77**: 145–58.

18 Guyatt AR, Kirkham AJ, Mariner DC, Baldry AG, Cumming G. Long-term effects of switching to cigarettes with lower tar and nicotine yields. *Psychopharmacology* 1989; **99**: 80–6.

19 Stephen A, Frost C, Thompson S, Wald N. Estimating the extent of compensatory smoking. In: Wald N, Froggatt P (eds). *Nicotine, smoking and the low tar programme.* Oxford: Oxford University Press, 1989: 100–15.

20 Peach H, Hayward DM, Shah D, Ellard GA. A double-blind randomized controlled trial of the effect of a low- versus a middle-tar cigarette on respiratory symptoms – a feasibility study. *IARC Sci Publ* 1986; **74**: 251–63.

21 Frost C, Fullerton FM, Stephen AM, Stone R, *et al.* The tar reduction study: randomised trial of the effect of cigarette tar yield reduction on compensatory smoking. *Thorax* 1995; **50**: 1038–43.

22 Lynch CJ, Benowitz NL. Spontaneous cigarette brand switching: consequences for nicotine and carbon monoxide exposure. *Am J Pub Health* 1987; **77**: 1191–4.

23 Peach H, Hayward DM, Ellard DR, Morris RW, Shah D. Phlegm production and lung function among cigarette smokers changing tar groups during the 1970s. *J Epidemiol Community Health* 1986; **40**: 110–6.

24 Waller RE, Froggatt P. Product modification. *Br Med Bull* 1996; **52**: 193–205.

25 Kozlowski LT, Mehta NY, Sweeney CT, Schwartz SS, *et al.* Filter ventilation and nicotine content of cigarettes from Canada, the United Kingdom, and the United States. *Tob Control* 1998; **7**: 369–75.

26 Gray N. Low-tar cigarettes: bane or benefit. *Cancer Detect Prev* 1987; **10**: 187–92.

27 National Cancer Institute Expert Committee. *The FTC cigarette test method for determining tar, nicotine and carbon monoxide yields of US cigarettes.* NIH Publication No. 96-4028. Smoking and Tobacco Control Monograph No. 7. Washington, DC: NCI, 1996.

28 Bross ID, Gibson R. Risks of lung cancer in smokers who switch to filter cigarettes. *Am J Pub Health Nations Health* 1968; **58**: 1396–403.

29 Wynder EL, Mabuchi K, Beattie EJ Jr. The epidemiology of lung cancer. Recent trends. *JAMA* 1970; **213**: 2221–8.

30 Wynder EL, Stellman SD. Impact of long-term filter cigarette usage on lung and larynx cancer risk: a case-control study. *J Natl Cancer Inst* 1979; **62**: 471–7.

31 Lubin JH, Blot WJ, Berrino F, Flamant R, *et al.* Patterns of lung cancer risk according to type of cigarette smoked. *Int J Cancer* 1984; **33**: 569–76.

32 Stellman SD, Garfinkel L. Lung cancer risk is proportional to cigarette tar yield: evidence from a prospective study. *Prev Med* 1989; **18**: 518–25.

33 Tang JL, Morris JK, Wald NJ, Hole D, *et al.* Mortality in relation to tar yield of cigarettes: a prospective study of four cohorts. *Br Med J* 1995; **311**: 1530–3.

34 Stellman SD. Cigarette yield and cancer risk: evidence from case-control and prospective studies. *IARC Sci Publ* 1986; **74**: 197–209.

35 Peto R. Overview of cancer time-trend studies in relation to changes in cigarette manufacture. In: Zaridze DG, Peto R (eds). *Tobacco: a major international health hazard.* IARC Scientific Publication No. 74. Lyon: IARC, 1986: 211–6.

36 Thun MJ, Day-Lally CA, Calle EE, Flanders WD, Heath CW. Excess mortality among cigarette smokers: changes in a 20-year interval. *Am J Pub Health* 1995; **85**: 1223–30.

37 La Vecchia C, Liati P, Decarli A, Negrello I, Franceschi S. Tar yields of cigarettes and the risk of oesophageal cancer. *Int J Cancer* 1986; **38**: 381–5.

38 La Vecchia C, Bidoli E, Barra S, D'Avanzo B, *et al.* Type of cigarettes and cancers of the upper digestive and respiratory tract. *Cancer Causes Control* 1990; **1**: 69–74.

39 Lopez Abente G, Gonzalez CA, Errezola M, Escolar A, *et al.* Tobacco smoke inhalation pattern, tobacco type, and bladder cancer in Spain. *Am J Epidemiol* 1991; **134**: 830–9.

40 Stellman SD. Influence of cigarette yield on risk of coronary heart disease and chronic obstructive pulmonary disease. *IARC Sci Publ* 1986; **74**: 237–49.

41 Parish S, Collins R, Peto R, Youngman L, *et al.* Cigarette smoking, tar yields, and non-fatal myocardial infarction: 14,000 cases and 32,000 controls in the United Kingdom. The International Studies of Infarct Survival (ISIS) Collaborators. *Br Med J* 1995; **311**: 471–7.

42 Participants of the Fourth Scarborough Conference on Preventive Medicine. Is there a future for lower-tar-yield cigarettes? *Lancet* 1985; **ii**: 1111–4.

43 Wynder EL, Goodman MT, Hoffmann D. Demographic aspects of the low-yield cigarette: considerations in the evaluation of health risk. *J Natl Cancer Inst* 1984; **72**: 817–22.

44 Jarvis MJ, Marsh A, Matheson J. Factors influencing choice of low-tar cigarettes. In: Wald N, Froggatt P (eds). *Nicotine, smoking and the low tar programme.* Oxford: Oxford University Press, 1989: 220–8.

45 Jarvis MJ, Wardle J. Social patterning of health behaviours: the case of cigarette smoking. In: Marmot M, Wilkinson R (eds). *Social determinants of health.* Oxford: Oxford University Press, 1999: 240–55.

46 Stellman SD, Garfinkel L. Smoking habits and tar levels in a new American Cancer Society prospective study of 1.2 million men and women. *J Natl Cancer Inst* 1986; **76**: 1057–63.

47 Jarvis MJ, Russell MAH. Sales-weighted tar, nicotine and carbon monoxide yields of U.K. cigarettes: 1985. *Br J Addict* 1986; **81**: 579–81.

48 Gori GB, Bock FG. *A safe cigarette?* Banbury Report 3. Cold Spring Harbor: Cold Spring Harbor Laboratory, 1980.

49 British American Tobacco. *Proceedings of the smoking behaviour-marketing conference July 9th-12th 1984.* Minnesota trial exhibit 13,431.

50 Hurt RD, Robertson CR. Prying open the door to the tobacco industry's secrets about nicotine: the Minnesota Tobacco Trial. *JAMA* 1998; **280**: 1173–81.50

7 | The management of nicotine addiction

7.1 General and non-pharmacological approaches
7.2 Nicotine replacement therapy
7.3 Non-nicotine medications for treating nicotine addiction
7.4 Evidence-based treatment of nicotine addiction
7.5 Nicotine replacement treatment in pregnancy
7.6 Cost-effectiveness of treating nicotine addiction

7.1 General and non-pharmacological approaches

There are two general approaches to the management of nicotine addiction, comprising:

1 Population interventions that can be delivered on a large scale, such as advertising campaigns, educational or self-help materials, products which can be bought over-the-counter, brief routine interventions by health care professionals, and telephone help-lines.

2 Individual interventions based on the traditional model of intensive expert-delivered treatments, and delivered to smokers either individually or in groups.

The two approaches overlap, and there are constant efforts to adapt for wider dissemination interventions proven to be effective in the intensive treatment setting.

In general terms, the greater the degree of contact between the smoker and the treatment provider in these interventions, the greater the efficacy achieved. However, it is also true that, in general, the more intensive the intervention the greater the financial cost and the smaller the proportion of the smoking population reached by the intervention. Treatment cost-efficacy is also clearly an important issue in this area because treatments have often been regarded as overly expensive for the results they deliver. This may not necessarily be the case since intensive interventions tend to attract the more highly dependent, and therefore high-risk, smokers.[1] However, even the most highly effective intervention service will have little impact on population smoking prevalence if only a minority of smokers use it, and it is the case that smokers in the UK seldom attend specialist treatment clinics. Even well run and centrally based smoking cessation clinics with large catchment areas can expect to attract only some 200–300 smokers per year. This is clearly a tiny minority of the smoking population, so the role of other interventions also needs to be assessed.

Comprehensive summaries of the existing evidence are available through the US Agency for Health Care Policy and Research guideline document[2] and the more recent Cochrane Centre reviews used in the preparation of evidence-based guidelines in the UK (see Section 7.4 for details).

Wide-reach approaches

In view of the scale of smoking prevalence, there is a pressing need for interventions reaching large populations. Numerous approaches have attempted to address this need and can be difficult to classify. There is a thin line dividing minimal interventions, which can still be considered treatments aimed at managing nicotine addiction (eg brief advice by a health professional), and educational or motivational approaches, which would not be defined as treatment (eg similar advice delivered through posters or mass media).

Comprehensive community-oriented interventions. A range of interventions can be targeted at whole communities (eg posters, mass-media campaigns, telephone and self-help interventions, 'quit and win' contests, etc). Such 'all-out' broad interventions have strong intuitive validity, but a number of sensible interventions have not in fact proven to be effective in the area of smoking cessation. Many community-level strategies have never been properly tested on their own, but a state-of-the-art comprehensive community level strategy has been carefully evaluated in the Community Intervention Trial for Smoking Cessation (COMMIT), the single biggest test of a smoking cessation intervention to date.[3,4] A community-level intervention comprising 58 mandated and numerous additional activities was tested in 11 matched pairs of communities in the US. The results were disappointing, showing no statistically significant effect on the target group of heavy smokers (>24 per day) or on total smoking prevalence, though there was a small effect of potential public health significance in light to moderate smokers. It has been pointed out that the same type of general and motivational interventions could well be more effective in countries less saturated with anti-smoking education.[5] Also, the lack of an overall effect does not necessarily mean that all the component parts were also ineffective; some of these were reaching only small groups of smokers, and may not have been run in an optimal way.

Self-help interventions. Self-help leaflets and books are an inexpensive way of distributing cessation advice to a large potential market. Most such materials, however, repeat variations of behavioural advice developed in the 1970s prior to full appreciation of the addictive nature of smoking and the advent of pharmacological treatments. From a contemporary point of view, the face validity of some of them is good, particularly when advice on nicotine replacement therapy (NRT) is included, but in others it is rather poor. An interesting development based on computing technology is to personalise advice to individual smokers.[6] In the Cochrane review of the relevant literature,[7,8] comprising 41 trials of variable methodological rigour, overall there was a small but significant effect of self-help in comparison with no intervention. There was also an advantage of personalised self-help materials in comparison with generic ones, but no incremental benefit of self-help materials when added to other brief interventions. The US guidelines[2] suggest that written materials may be more effective when combined with other more intensive therapies.

Telephone help-lines. When well advertised, telephone help-lines can elicit a large number of calls.[9] Although there is no clear evidence on the effectiveness of brief outreach telephone counselling, a large study of up to six proactive 50-minute counselling sessions with smokers ready to quit found this intervention to be effective.[10] Clear research evidence is lacking for the more usual responsive telephone counselling but, even in the absence of strong evidence of efficacy, existing responsive help-lines such as Quitline in the UK can assume another potentially important role in referring callers to locally available face-to-face treatments.

Advice by physicians. A review of 31 randomised controlled trials from both primary and hospital care concluded that brief advice by doctors is effective, and that more intensive interventions only marginally increase the efficacy of brief advice.[11] Overall, the efficacy is low (1–3% above controls), but because of wide reach this approach has the potential to influence smoking prevalence in whole populations. It should be noted that, as might be expected, brief interventions are effective primarily with light smokers.[12] Brief interventions by other health care staff are also effective, to a degree similar to that achieved by clinicians.[1]

Despite its proven value, it seems difficult to initiate and maintain routine smoking cessation interventions in general practice with all smokers. The low success rates can be demoralising, insistent advice can strain the doctor-patient relationship, general practitioner (GP) consultations often have other pressing priorities, and the time available for extra input is short.[13]

Intensive interventions

Behavioural interventions. Smoking cessation clinics in the UK typically offer a combination of NRT and behavioural support delivered over a series of weekly sessions. Both individual and group behavioural treatments are effective.[7,8,14] The Cochrane reviews found no difference in efficacy between individual and group treatments, so preference must be given to group approaches where practicable as they are considerably more cost-effective, though some smokers prefer the individual approach. The large Lung Health Study[15] provides a persuasive confirmation of the effectiveness of intensive support combined with NRT, achieving as it did a one-year abstinence rate of 35% versus 9% with usual care (see Section 5.2 for details). Two aspects of intensive behavioural treatments deserve a separate note:

- *Relapse prevention.* Well-run intensive treatments are very effective in helping even heavy and dependent smokers to stop smoking for a period of several weeks. As with other addictions, however, relapse is the major problem, eroding the success rates of 50–60% at one month to 20–30% one year later.[16] This is still a respectable success rate because, without treatment, cessation rates are very low. Nevertheless, the lack of effective relapse prevention is felt acutely in clinical practice. Several relapse prevention procedures have been proposed and are widely practised, but so far none has proven clearly effective.
- *Social support.* Social support is believed to be an important ingredient of intensive treatments, and of group treatments in particular. Having another person or a group of fellow smokers involved in a quit attempt seems to provide a boost to the effort to succeed. A recent trial of an attempt to translate this mechanism of intervention to general practice by 'buddying' pairs of smokers showed a significant short-term effect in comparison with a control group.[17] More work is needed on designing and evaluating social support treatments applicable on a larger scale.

Hypnosis and acupuncture. Hypnosis and acupuncture hold a special place among a range of existing treatments for smokers. While most other treatments are practised only on a very limited scale, the number of commercial advertisements for acupuncture and hypnosis suggests that these two approaches are popular with smokers. The Cochrane group has reviewed nine studies of hypnosis[18] and 16 trials of acupuncture,[19] and concluded that evidence of specific efficacy is lacking. Some people can of course be helped by numerous different unproven procedures, via non-specific and placebo effects.

7.2 Nicotine replacement therapy

NRT is the only pharmacotherapy licensed in the UK to manage nicotine addiction. At the time of writing, there are five different NRT products on the market, some produced in different strengths and different versions. This section outlines their general features, as well as their individual characteristics, with special attention to issues of efficacy, dose-response, dependence potential, NRT combinations, and choosing the right product for individual smokers.

Mode of action

The main short-term difficulties smokers experience when trying to stop smoking seem to be attributable to acute nicotine withdrawal. The basic idea behind using nicotine replacement is to break the quitting process into two phases. In the first phase, smokers learn to cope without smoking behaviour and regular rapid boli of nicotine, while protected from the worst withdrawal effects by moderate levels of nicotine provided by NRT. Later, nicotine is gradually withdrawn completely.

NRT alleviates withdrawal discomfort.[20] However, as has been pointed out, the severity of withdrawal is only a weak predictor of success in stopping smoking. Although such alleviation probably constitutes the main effect of NRT, other mechanisms may also have a role, such as the provision of a coping mechanism, or even the replacement of some of the hypothetical positive effects of nicotine.[21] NRT may also make early relapses to smoking in smokers trying to quit less rewarding, and therefore less likely to trigger a full-scale relapse. A related possible mechanism could be deconditioning because the link between pharmacological reinforcement and smoking behaviour may weaken during abstinence accompanied by NRT use.

Overall efficacy

Whatever the actual mechanism, there is ample evidence that NRT is effective in helping smokers quit. It has been by far the most extensively and rigorously tested smoking cessation treatment. The most recent update of the Cochrane review (May 1999)[22] summarises 49 trials of nicotine gum, 24 of transdermal nicotine patch, four of nicotine spray and four of nicotine inhalators. The overall odds ratio for abstinence with NRT compared to placebo was 1.73 (95% confidence interval 1.60–1.86). These odds were maintained regardless of the intensity of additional support or the setting in which the NRT was offered. In addition to enhancing early cessation, there is evidence that NRT also reduces early relapse.[23]

Dose response

Increasing nicotine dose seems to increase treatment efficacy, though the dose-response curve is shallow.[24] For nicotine chewing gum, the 4 mg gum was shown to be more effective than the 2 mg gum in highly dependent smokers, though it is not known whether this also applies to less dependent smokers. A recent large-scale trial of nicotine transdermal patches (the Collaborative European Anti-Smoking Evaluation (CEASE) trial) has shown a small, but significant advantage of 25 mg over 15 mg patches in sustained one-year abstinence rates.[25] Some other trials have also shown higher delivery patches to be more effective,[26,27] though in one large study the effect did not reach significance,[24] and in another substantial study it was absent.[28]

Given the apparently stronger effect of higher nicotine doses with a self-dosing oral preparation such as the gum, increasing the nicotine delivery of these products should be further examined.

Dependence potential

Fears have been expressed about the dependence potential of NRT since it first appeared on the market. Among smokers' clinic patients, some 25% of long-term successes who started on the nicotine gum were still using it at one year,[29] as were 43% of continuing long-term successes of a similar sample who started on nasal spray.[30] No cases of long-term use of the nicotine patches have been reported, suggesting that the incidence of long-term use may be related to the speed of nicotine absorption from individual products. The nasal spray study[30] allowed the monitoring of the effect on relapse rate of withdrawing the medication after one year. Reassuringly, there was no difference in relapse rate between those who used the spray for one year and those who did not.[31] Long-term NRT users seem closer in their characteristics to treatment failures than to 'NRT-free' long-term treatment successes, and it is likely that many of them would be smoking if NRT was not available over a protracted period of time.[29] NRT does not create new dependence as users are already dependent on nicotine, of which NRT typically provides less than cigarettes and in a much safer way.[32] Long-term use of NRT does not seem to be associated with any significant harmful effects.[15].

Additional support

Current moves to make NRT treatments more widely available have the potential to bring about substantial public health benefits.[33] There is a reservation that without 'psychological packaging', involving explanation of effects and creation of adequate expectations, smokers may not adequately engage with treatment and NRT may then be less effective. Appropriate briefing of smokers purchasing NRT, through well designed leaflets, or even via repeated mailings of tailored product and behavioural advice, should be able to offset this effect.

Currently available nicotine replacement therapy products

Nicotine chewing gum. Three brands of nicotine chewing gum are on the market, all in 2 mg and 4 mg doses, and all equivalent in their nicotine delivery. Smokers are instructed to chew each piece of gum slowly for some 30 minutes. The usual recommendation is to chew one piece per hour, supplemented by opportunistic use as required, with total usage of up to 15 pieces a day. Nicotine enters the bloodstream via the buccal mucosa, and the plateau blood nicotine level is reached after about 30 minutes. Only about 0.9 mg nicotine from one piece of 2 mg gum and 1.2 mg from 4 mg gum reaches the bloodstream.[34]

Nicotine transdermal patch. There are four brands of patch on the UK market. The highest doses of all preparations aim to deliver about 1 mg of nicotine per hour. Some patches are designed to be worn for only 16 hours to avoid sleep-time nicotine dosing, while 24-hour patches are aimed at prevention of urges to smoke on waking. Nicotine absorption is slow, taking hours to reach a plateau. Users have no control over the nicotine dose but, because of simple instructions and ease of use, compliance with patch use is much better than with other NRT products.[35]

All existing types of patch come in three different strengths, to be used in similar schedules, starting with the strongest dose and using the weaning-off doses for up to three months. Eight weeks of patch use has been shown to be as effective as longer courses of treatment, and there is no evidence that tapered use is better than abrupt withdrawal.[22,25]

Nicotine nasal spray. Nicotine nasal spray consists of a bottle of nicotine solution which is sprayed into a nostril by an air pump plunger via a nozzle. It provides by far the most rapid nicotine absorption among the NRT products, reaching a plateau in about 10 minutes. A single spray into one nostril delivers about 0.5 mg of nicotine, absorbed mostly through the nasal mucosa. The recommended usage is one spray in each nostril hourly up to 16 times a day. There is some evidence that the spray is especially helpful to more highly dependent smokers.[30] The spray is initially unpleasant to use. Although users who persevere get used to the irritant effects, this is a serious drawback and necessitates close initial supervision and reassurance of first-time users.

Nicotine inhalator. The inhalator consists of a plastic holder resembling a cigarette holder, and cartridges containing a polythene plug impregnated with nicotine. Puffing on the inhalator brings nicotine vapour into the mouth and throat, where it is absorbed. The nicotine does not reach the lungs. About 20 puffs on the inhalator equal one puff on a cigarette, so frequent and intensive puffing for some 20 minutes is needed to obtain about 1 mg of nicotine. Users are advised to use 6–12 cartridges per day, each of them for three puffing sessions. The inhalator has some appeal in promising to replace some of the behavioural aspects of smoking, but users tend to feel self-conscious in public.[35] This may change with wider awareness of the device.

Nicotine sublingual tablet. Nicotine tablets are designed to be held under the tongue until they dissolve, usually within 20–30 minutes. The nicotine absorption profile and the dose delivered from the preparation are similar to that of nicotine chewing gum and nicotine inhalator. These formulations, particularly if used heavily, may be more user friendly than the gum, though experience with these relatively new products is currently limited.

Differences between products in efficacy

There is no evidence that, overall, products with similar nicotine delivery differ in efficacy.[35] The one study that directly compared 16-hour and 24-hour patches found no difference in efficacy.[36] It remains a possibility that different products may differ in acceptability, the adherence of users to recommended usage, and ultimately in effects on outcome for special subgroups of clients.

Nicotine replacement therapy combinations

There is a growing interest in evaluating the notion that treatment efficacy may be increased by using two or more NRT products at the same time. This could arise via increased nicotine intake and, in the case of combinations including nicotine patches, via supplementing the steady nicotine levels from the patch with occasional boosts from the faster delivery products to counteract surges in withdrawal discomfort.

The evidence supporting product combinations has so far been rather weak. The Cochrane review concluded that there is no strong evidence that NRT combinations are more effective than single products.[22] Other studies have been published since the last update of the Cochrane review. For example, adding the patch to nicotine inhaler[37] or nicotine nasal spray[38] did not improve mid-term efficacy of the inhaler or spray alone. One study reported a combination of spray and patch to be superior to patch alone, but did not include a spray-only treatment group.[39]

7.3 Non-nicotine medications for treating nicotine addiction

NRT medications are effective and safe,[40] but some smokers prefer not to use them because they are sceptical of the replacement rationale, or fear addiction, or believe that nicotine is harmful. In addition, many smokers have tried NRT, but have not succeeded in quitting[41] and might benefit from switching to a non-nicotine medication.

The following paragraphs briefly review non-nicotine medications that have been tested. Placebo-controlled trials are emphasised. Data from non-humans and on non-cessation outcomes (relief of withdrawal symptoms or suppression of post-cessation weight gain) are not discussed. Recent reviews give more detailed information.[42–45]

Antidepressants

One common rationale for using antidepressants for smoking cessation has been that many smokers either have depressive symptoms or develop them post-cessation, and that these interfere with cessation.[46] Another rationale is that nicotine dependence is associated with low dopamine levels which are corrected by some antidepressants.[47]

Bupropion. Currently, bupropion is the only non-nicotine therapy for smoking cessation marketed in the USA.[47,48] At the time of writing, it is likely to become available for use in the UK in the near future. Bupropion is an atypical antidepressant with both dopaminergic and adrenergic actions. Treatment begins one week prior to cessation, and lasts 7–12 weeks. In four trials, bupropion, like NRT, doubled quit rates.[48] Importantly, its efficacy does not appear to be due to its classical antidepressant effects because the drug works equally well in smokers with and without a past history of depression. Like NRT, bupropion also appears to reduce post-cessation weight gain,[47] at least while it is being used. One study has reported that combined nicotine patch and bupropion produced higher quit rates than the nicotine patch alone.[49] Side effects are mild and consist of nausea and insomnia. Earlier trials in depressed patients suggested that bupropion increased the risk of seizures, but more recent data with the slow-release preparation indicate that with doses of 300 mg per day or less, and with simple screening, the risk of seizures is small and no more than with other antidepressants.[47–49]

Nortriptyline. The other antidepressant that appears to increase cessation is nortriptyline, a tricyclic antidepressant with mostly noradrenergic properties and little dopaminergic activity. Two published trials indicate that nortriptyline increases cessation rates, an effect apparently unrelated to depressive symptoms.[50,51] Side effects from nortriptyline include anticholinergic effects, nausea and sedation.

Other antidepressants. In one study, imipramine, which has mostly noradrenergic and serotonergic effects, did not improve smoking cessation,[52] but a small study suggested that doxepin, which has similar neurochemical effects, increased short-term cessation.[53]

Cigarette smoking inhibits the enzyme monoamine oxidase (MAO)-A.[53–55] This enzyme breaks down acetylcholine, so smoking increases cholinergic and adrenergic transmission. Moclobemide is an MAO inhibitor which has been used in smoking cessation to replace the MAO inhibition of smoking and, in fact, was effective in the short term in the one published study.[56]

Of the selective serotonin reuptake inhibitor group of antidepressants, fluoxetine has been tested in two large multicentre trials, one of which went unreported. The other study reported increased cessation with fluoxetine, but did not report abstinence rates.[57] One study of venlafaxine reported negative results.[58]

In summary, there is clear evidence that bupropion is an effective non-nicotine therapy, and suggestive evidence that nortriptyline may also be effective. Across antidepressant studies, those with adrenergic activity appear to be more likely to be effective than those without it. Thus, although many workers have focused on dopaminergic activity to explain antidepressant efficacy for smoking cessation, it may be noradrenergic activity that is responsible for efficacy.

Clonidine

Clonidine is an α2-noradrenergic agonist that suppresses sympathetic activity; it is mostly used for hypertension, but also to reduce alcohol and opiate withdrawal.[59] Both as pills and as a patch in low doses (usually 0.2–0.4 mg per day), clonidine increased smoking cessation in eight of nine trials.[59] Although early studies suggested that clonidine was effective only for women, later studies have not found this. Clonidine has more significant side effects (eg sedation, postural hypotension) and more drop-outs due to side effects than NRT. For these reasons, it is mainly used as a second-line drug for those who cannot, or do not wish to, take NRT or bupropion.

Mecamylamine

Mecamylamine is a nicotine antagonist originally used to decrease cholinergic activity, and thus reduce blood pressure.[60,61] It does not specifically bind at the nicotinic receptor but blocks the associated ion channel; this may be why, in humans, it blocks the effects of nicotine but does not precipitate withdrawal symptoms.[60,61] Early trials with moderate-level doses of mecamylamine alone produced unacceptable side effects. Based on a theoretical analysis, two trials have tested a combination of nicotine patch with low doses of mecamylamine (5–10 mg per day), and both found that this combination was superior to placebo.[62] In the later study, mecamylamine alone was helpful early on, but not at later follow-up. However, the preliminary report of an even more recent multicentre trial failed to confirm efficacy.[63] Although some gastrointestinal effects, particularly constipation, were significant, side effects with the lower doses and with the counteracting effects of nicotine were not a major cause of drop-out.

Buspirone

Buspirone is a serotonin agonist which acts as a non-sedating, non-addicting anxiolytic. Buspirone has been tested as an aid to stopping smoking because anxiety is commonly associated with cessation. The first study found a significant effect on short-term cessation in normal (non-anxious) smokers.[64] A second study found that buspirone increased quit rates among smokers with above average pre-cessation anxiety levels,[65] while a third found buspirone to be ineffective in both low- and high-anxiety smokers.[66] Ondansetron, a serotonin antagonist, failed to improve cessation in one study.[67]

Lobeline

Lobeline is a nicotine-related alkaloid with some, but limited, activity at nicotinic receptors,[68] and is found in several over-the-counter smoking cessation aids. Both older studies[69] and two more recent studies[70,71] have failed to find that lobeline increases quit rates.

Naltrexone

Naltrexone (a long-acting formulation of naloxone) blocks opioid release, and is used to treat both opioid and alcohol dependence. Two randomised trials reported positive short-term results,[71,72] but a recent trial failed to confirm efficacy.[73]

Silver acetate

Silver acetate, found in some over-the-counter preparations, interacts with cigarette smoke to produce an aversive metallic taste. Both older[74] and newer[75] studies with silver acetate have failed to find evidence of efficacy.

Sensory replacement

Although often overlooked, sensory effects of smoking are important to its reinforcing effects, and their absence during cessation may be a cause for relapse.[76] Three clinical trials of inhalers with chemicals to mimic the sensory feel of cigarettes have been tested:

- an ascorbic acid aerosol increased short-term cessation[77]
- a citric acid inhaler increased very short-term cessation but only in heavy smokers,[78] and
- a citric acid inhaler plus nicotine patch increased short-term cessation over a placebo inhaler plus nicotine patch.[79]

Sensory replacement therapy could be useful for the many smokers who report missing the sensory aspects of smoking, for the minority of smokers for whom nicotine dependence is not an important barrier to cessation or as an adjunct to NRT or non-nicotine therapies. Thus, further testing appears warranted.

Summary

Bupropion and clonidine are the only proven non-nicotine therapies for smoking cessation (see Table 7.1). Of these, bupropion is preferred because of its better side effect profile. Nortriptyline, moclobemide, mecamylamine and sensory replacement have shown preliminary promise.

Table 7.1. Non-nicotine therapies for smoking cessation.

Proven therapies:	Bupropion Clonidine
Possible therapies:	Nortriptyline Noradrenergic antidepressants MAOIs Mecamylamine + NRT Sensory replacement
Therapies that have been tested but found to be ineffective, or that at the present time lack sufficient evidence:	Anorectics Benzodiazepines Beta-blockers Buspirone Caffeine/ephedrine Cimetidine Dextrose Lobeline Naltrexone Odansetron Phenylpropanolamine Silver acetate Stimulants SSRI antidepressants

MAOI = monoamine oxidase inhibitor; NRT = nicotine replacement therapy; SSRI = selective serotonin reuptake inhibitor.

Bupropion has been widely used in the USA and appears to be popular.[41] This could be because it is the first non-nicotine therapy to become widely available to smokers, or because many smokers have failed NRT and are looking for new treatments. In either case, it suggests that there is a future for non-nicotine therapies for smoking cessation, and that smoking cessation services need to be able to adapt to deliver bupropion and other new developments that are shown to be effective and cost-effective.

7.4 Evidence-based treatment of nicotine addiction

Nicotine addiction occupies a strange position in the health care system of Britain, and probably of most countries. Treatment of nicotine addiction has been shown to be both effective[80] and cost-effective;[81] despite this, at the time of writing it is still not universally available as part of the NHS. This situation is particularly inconsistent, given that the NHS provides treatment for addicts to illicit drugs and to alcohol. A 1995 Health Education Authority (HEA) survey, *Health in England*,[82] found that only about 29% of smokers who had seen their GP in the previous year said they had been given advice on smoking. A 1996

HEA survey[83] reported that only 39% of pregnant smokers said they had received advice about smoking. The limited evidence available on the delivery of treatment through the NHS shows that it is variable[84] and depends on the enthusiasm and commitment of individuals.

The barriers to action amongst GPs include lack of time, perceived lack of skills, and the perception that success rates are low.[2,85,86] The last of these may be particularly important in engaging the interest of GPs. Appreciating the difference between success rates and the numbers reached may help: intensive treatments that achieve high cessation rates but reach limited numbers will usually produce fewer ex-smokers than less intensive approaches which reach many smokers. Thus, brief advice from GPs (defined as up to 3 minutes) (see Table 7.2) may encourage 'only' about 2% more smokers to stop compared with normal care control, but this apparently low figure applied nationally to all GPs would be enormously worthwhile and cost-effective. With

Table 7.2. Estimates of the effect of smoking cessation interventions on smoking abstinence rates at six months (*source:* Ref 80).

Intervention element	Data source	Increase in % of smokers abstinent for 6 months or longer
Very brief advice to stop (3 min) by clinician, versus no advice	AHCPR[2]	2
Brief advice to stop (up to 10 min) by clinician, versus no advice	AHCPR[2]	3
Adding NRT to brief advice versus brief advice alone or brief advice plus placebo	Cochrane[22]	6
Intensive support (eg smokers' clinic) versus no intervention	AHCPR[2]	8
Intensive support plus NRT versus intensive support or intensive support plus placebo	Cochrane[22]	8
Cessation advice and support for hospital patients versus no support	AHCPR[2]	5
Cessation advice and support for pregnant smokers versus usual care or no intervention	AHCPR[2]	7

Note: to estimate the overall effect of a particular package of treatment (eg intensive behavioural support plus nicotine replacement therapy (NRT)), broadly speaking, the effects of the elements can be added together. Thus, intensive support plus NRT can increase long-term abstinence rates by some 16% (8% intensive support plus 8% NRT) over control.
AHCPR = US Agency for Health Care Policy and Research.

an average of about 9,000 patients and 2,600 smokers (29%) in a five-partner practice, brief advice could help more than 50 (2% of 2,600) additional smokers each year in that practice to stop. Nationally, this would produce about 300,000 additional ex-smokers. With NRT added to usual care (see Table 7.2), the result would be about 156 (6% of 2,600) extra ex-smokers per practice, or almost one million nationally. These apparently low absolute figures are worthwhile and extremely cost-effective compared to many other things doctors do,[81] and this message needs to be conveyed clearly to health professionals.

The relative lack of progress in the last 20 years clearly indicates that education and persuasion are not enough. Until structural factors are addressed, including renegotiating the core contracts of relevant health professionals to include smoking cessation treatment, the situation will probably not change. In effect, a change is needed in the culture of the NHS towards the delivery of effective preventive interventions. Such a change will not come easily, and will require the support of government as well as of health professionals.

This may be beginning to happen in England. In December 1998, the government launched a White Paper on tobacco control,[86] the first ever in the UK, which for the first time set out a framework for an NHS smoking cessation treatment service. Funding was allocated to develop these services, initially in relatively deprived areas called health action zones (HAZ) for one year, with the promise of money for two additional years and more national coverage if all goes well in the first year. The treatment services envisaged are essentially those described in the evidence-based guidelines published in Britain at about the same time as the White Paper.[80,81] The recommendations of the guidelines are summarised in Tables 7.2 and 7.3. Key features of the guidelines are that they:

- are evidence-based
- are endorsed by the professions, and
- now have some financial support from central government to begin establishing treatment services.

It remains to be seen whether the changes recommended in the guidelines will be widely adopted into the fabric of the health service.

7.5 Nicotine replacement treatment in pregnancy

There is little information available on the use and relative risks of NRT in pregnancy, so it is difficult to formulate policy on the use of NRT in this indication.[80,85,86] However, cigarette smoking, in general, delivers more nicotine and results in higher levels of organ exposure

Table 7.3. Recommendations for health professionals.[80]

Recommendations for all health professionals
- Assess the smoking status of patients at every opportunity; advise all smokers to stop; assist those interested in doing so; refer to specialist cessation service if necessary; recommend smokers who want to stop to use NRT; provide accurate information and advice on NRT.
- Smoking and smoking cessation should be part of the core curriculum of the basic training of all health professionals.

Recommendations for the primary care team
- Assess the smoking status of patients at every opportunity; advise all smokers to stop; assist those interested in doing so; offer follow-up; refer to specialist cessation service if necessary; recommend smokers who want to stop to use NRT; provide accurate information and advice on NRT.

Recommendations for smoking cessation specialists
- Intensive smoking cessation support should, where possible, be conducted in groups, include coping skills training and social support, and should offer around 5 sessions of about 1 hour over about 1 month, and follow-up.
- Intensive smoking cessation support should include the offer of or encouragement to use NRT, and clear advice and instruction on how to use it.

Nicotine replacement therapy
- Smokers should be encouraged to use NRT as a cessation aid. It is effective and safe if used correctly.
- Health professionals who deliver smoking cessation interventions should give smokers accurate information and advice on NRT.
- Consideration should be given to ways of increasing the availability of NRT to low income smokers, including at reduced cost or free of charge.

Recommendations for specific smoker populations
- Hospital staff should assess the smoking status of patients on admission, advise smokers to stop, and assist those interested in doing so; patients should be advised of the hospital's smoke-free status before admission.
- Hospital patients who smoke should be offered help in stopping smoking, including the provision of NRT.
- Pregnant smokers should be given firm and clear advice to stop smoking throughout pregnancy, and given assistance when it is requested.
- Cessation interventions shown to be effective with adults should be considered for use with young people, with the content modified as necessary.

Recommendations for health commissioners
- To produce cost-effective significant health gain in the population, smoking cessation interventions should be commissioned.
- Review current practice, identify needs, and provide core funding to integrate smoking cessation into health services; plan a cessation strategy with public health specialists; seek advice from smoking cessation specialists.
- These plans should include a specialist cessation service.
- Training should be a core part of a smoking cessation programme in all health authorities; protected time and funding should be built into this programme.
- Core fund smoking cessation training, or make sure that smoking cessation is prioritised within existing training budgets.
- Make provision to ensure that NRT is available to hospital patients who need it, in conjunction with professional advice and cessation support.
- Require all services, departments and clinics to introduce systems to maintain an up-to-date record of the smoking status of all patients in their (paper or electronic) notes; it should be regarded as a vital sign.
- Ensure that all health care premises and their immediate surrounds are smoke-free.
- Work with clinicians to put systems in place to audit smoking cessation interventions throughout the health care system.

NRT = nicotine replacement therapy

than NRT, and also exposes the individual to many other toxins (see Section 2.6). It therefore seems likely that, whilst smoking cessation through non-pharmacological means remains the ideal, nicotine replacement is likely to be appreciably safer to the mother and fetus than continued smoking. NRT is therefore theoretically justified in pregnant women in whom non-pharmacological interventions have failed. However, because of concerns about excessive exposure to nicotine over time, it may be prudent to use shorter-acting NRT products such as gum or lozenges rather than patches. It also makes sense to attempt to dose NRT so as not to exceed the levels of nicotine in the body derived from cigarette smoking for an individual pregnant smoker.

Research into the use of NRT in pregnancy presents a number of ethical problems but is urgently needed, particularly in relation to heavy smokers in whom non-pharmacological behavioural therapies have failed.

7.6 Cost-effectiveness of treating nicotine addiction

There is clear evidence, summarised above, that a range of interventions for the treatment of nicotine addiction are effective. However, health resources are scarce, and there are many competing demands on NHS funds, so it is also important to consider the cost-effectiveness of treatments for nicotine addiction. Such studies combine data on the costs of treatments with evidence of effectiveness.

Synthesising evidence on cost-effectiveness is problematic. Relatively little resource information has been obtained concurrently with clinical trials, and most has been estimated retrospectively. Also, even if better resource data were available, the costs of different means of delivering interventions vary across countries and over time, complicating comparisons of results across studies, plus the fact that existing studies have included different costs and consequences. Cheung and Tsevat[87] criticised many published studies for excluding the consequences of changing smoking rates on health service use. The costs and consequences considered depend on the perspective taken. This is an important issue for NRT in the UK. Currently, most NRT therapy is bought from pharmacies by individuals attempting to quit, and the cost of these products falls therefore to the individual rather than to the health service. From a narrow health service perspective, these therapies currently appear to have low direct health service costs. However, from a wider societal viewpoint, the costs to individuals, whether for products or for the time involved in other therapies, would be included as part of the overall analysis.

Another important issue which influences the results from cost-effectiveness analysis is the comparison of the alternative interventions being considered. Most studies have suggested that both NRT therapies and brief advice yield value for money compared to other health care interventions when applied to the whole population of smokers. These results suggest that investing any additional resources in smoking cessation – or indeed switching from some other health care interventions into smoking cessation interventions – would be worthwhile.

There is less agreement about the most efficient means of spending any allocated budget for smoking cessation. A review of cost-effectiveness studies up to 1997 suggested that it could be concluded that the least intensive interventions, such as brief advice or the use of self-help manuals, yielded more favourable cost-effective ratios than some more intensive therapies.[88] For the UK, this conclusion depends crucially on the perspective taken,[89] since in this analysis NRT compared favourably with advice alone when only the NHS costs were considered.

Cromwell et al[1] investigated the cost-effectiveness of American guidelines covering a range of smoking interventions, and concluded that the more intensive the intervention, the lower the cost per quality adjusted life-year (QALY) saved. However, this study included more intensive interventions than had been previously considered, consisting of five individual or seven group counselling sessions with or without NRT. Five main types of interventions were considered both without NRT and with either patches or gum, making 15 intervention types in all:

- minimal counselling
- brief counselling
- full counselling
- individual intensive counselling, and
- group intensive counselling.

The cost-effectiveness figures were calculated, first assuming that 75% of American smokers wanting to quit had one of these interventions compared to no interventions being available. The interventions with patches were found to be more cost-effective than the equivalent intervention without NRT and with NRT gum, apart from intensive group counselling where the figures were almost equivalent for patches and no NRT. It is not clear why these findings differ from previous findings.

More realistically, the authors calculated a scenario in which patient preferences for different interventions were used. This resulted in an estimate of $1,915 cost per QALY across the whole smoking population. This figure compares favourably with the cost per QALY from the more unrealistic scenario of everyone being offered only one type of intensive intervention. Clearly, the overall financial costs of the combined intervention package would be lower.

More intensive interventions were also costed in an analysis of data for Britain.[81] In this model, a more stepped approach was assumed, with some smokers receiving brief advice alone and others – possibly those who had failed previously – being offered the more intensive interventions. Different stepped programmes were assessed, comprising:

- brief advice alone
- brief advice plus self-help material
- brief advice, self-help and NRT (for a proportion of smokers), and
- brief advice, self-help, NRT and specialist cessation interventions.

Given the current low level of such intensive services, it was assumed that only 2% of smokers would wish to use such interventions. Summary results from that study are shown in Table 7.4. From an NHS perspective, the different programmes, assumed to be directed at 50% of current smokers in any year, yield comparable cost-effectiveness figures when compared to the existing level of services. The costs to the NHS from implementing the programme with the intensive elements, based on the smoking cessation guidelines for health professionals,[80] were estimated at £331,000 per health authority.

Stapleton et al[90] also present some recent UK estimates based on resource data concurrently collected in a randomised controlled trial. The study was based on interventions provided within primary care, and presents the figures for the incremental (extra) costs and life-years

Table 7.4. Cost-effectiveness of smoking cessation programmes, England and Wales, 1996 prices (source: Ref 81).

	Cost per life year saved (£)	
	NHS perspective	Societal perspective
Brief advice	174	212
Brief advice + self-help	221	259
Brief advice + self-help + NRT	269	606
Brief advice + self-help + NRT + specialist cessation service	255	873

Notes:
- All interventions compared to current smoking cessation practice assuming additional resources to reach 50% of smokers per year.
- Numbers receiving different interventions based on expert opinion and available data.
- All health gains discounted at 1.5%.
NRT = nicotine replacement therapy.

saved of counselling with nicotine patch treatment over GP counselling alone. All interventions were directed at the more dependent
smokers, and NRT was provided free. The authors used the trial results
to simulate a situation in which NRT was available through prescriptions, implying that the NHS would bear more of the costs of these
types of interventions than currently. However, the costs of counselling
and prescribing NRT were based on the optimal cost-effectiveness
assumption that such services would be provided only if participants
remained abstinent. For abstinent smokers, repeat prescriptions for
NRT were made available for up to 12 weeks. From an NHS perspective, the results suggest that the cost per life-year gained ranged from
£345 for those aged 35–44 to £785 for those aged 55–65.

Currently, however, NRT products are not generally available on
prescription in the UK. Most estimates of cost-effectiveness have taken
figures from trials where the products were provided free of charge.
There is evidence to suggest that the take-up of NRT products will
vary depending on the charges. For example, Curry et al[91] found a
positive relationship between take-up of NRT and insurance cover in
the USA, with more cover increasing the cost-effectiveness of interventions. More evidence should become available from the areas where
NRT has been made available free for one week as part of the HAZ
initiative arising from the recent White Paper.[86]

Most economic studies of smoking cessation have been based on
the whole population of smokers, although some (such as the study by
Stapleton et al[90]) have presented results differentiated by age. Special
subgroups may yield even higher gains. There are no UK studies, for
example, on the cost-effectiveness of interventions with pregnant
women, but data from the US suggests that such interventions lead to
immediate cost savings for health authorities, quite apart from the
health gains for mothers and their children.[89] Similar savings may be
apparent for other high-risk groups. Lightwood and Glantz[92] calculated the health cost savings from reducing the incidence of acute
myocardial infarction (MI) and strokes in the USA. Krumholz et al[93]
estimated a cost of $220 per life-year saved from a smoking cessation
programme directed at those recovering from an MI, but that figure
excludes any estimate of health care cost savings.

Conclusions

In conclusion, all the available evidence suggests that treatment for
nicotine addiction can be obtained at a very low cost per life-year
gained compared to most other health care interventions.[94,95]
Investment in treatment for smokers would yield considerable health

gains at low cost. All the costs per life-year saved presented in this section are well below the informal figure of £6,000 per incremental life-year gained which was previously used as an initial yardstick by the Department of Health when considering whether the NHS should adopt new therapies or techniques. It is likely that few of the general health care interventions in current and future use in Britain will have a lower cost per life-year saved or QALY than the most pessimistic estimates of the cost-effectiveness of smoking cessation intervention.

References

1   Cromwell J, Bartosch W, Fiore M, Hasselband V, Baker T. Cost-effectiveness of the clinical practice recommendations in the AHCPR guideline for smoking cessation. *JAMA* 1997; **278**: 1759–66.

2   Fiore MC, Bailey WC, Cohen SJ, Dorfman SF, *et al. Smoking cessation: clinical practice guideline No. 18.* Rockville, MD: US Department of Health and Human Services, Agency for Health Care Policy and Research, 1996.

3   The COMMIT Research Group. Community intervention trial for smoking cessation (COMMIT): I. Cohort results from a four-year community intervention. *Am J Pub Health* 1995; **85**: 183–92.

4   The COMMIT Research Group. Community intervention trial for smoking cessation (COMMIT): II. Changes in adult cigarette smoking prevalence. *Am J Pub Health* 1995; **85**: 193–200.

5   Foulds J. Strategies for smoking cessation. *Br Med Bull* 1996; **52**: 1–17.

6   Strecher V, Kreuter M, Den Boer D, Kobrin S, *et al.* The effects of computer-tailored smoking cessation messages in family practice settings. *J Fam Pract* 1994; **39**: 262–70.

7   Lancaster T, Stead LF. *Individual behavioural counselling for smoking cessation (Cochrane review).* In: The Cochrane Library, Issue 2, 1999. Oxford: Update Software.

8   Lancaster T, Stead LF. *Self-help interventions for smoking cessation (Cochrane review).* In: The Cochrane Library, Issue 2, 1999. Oxford: Update Software.

9   Platt S, Tannahill A, Watson J, Fraser E. Effectiveness of antismoking telephone helpline: follow up survey. *Br Med J* 1997; **314**: 1371–5.

10 Zhu S, Stretch V, Balabanis M, Rosbrook B, *et al.* Telephone counseling for smoking cessation: effects of single-session and multiple session interventions. *J Consult Clin Psychol* 1996; **64**: 202–11.

11 Silagy C, Ketteridge S. *Physician advice for smoking cessation (Cochrane review).* In: The Cochrane Library, Issue 2, 1999. Oxford: Update Software.

12 Jackson P, Stapleton J, Russell M, Merriman R. Predictors of outcome in a general practitioner intervention against smoking. *Prev Med* 1986; **5**: 244–53.

13 Coleman T, Wilson A. Anti-smoking advice in general practice consultations: general practitioners' attitudes, reported practice and perceived problems. *Br J Gen Pract* 1996; **46**: 87–91.

14 Stead LF, Lancaster T. *Group behaviour therapy programmes for smoking cessation (Cochrane review).* In: The Cochrane Library, Issue 2, 1999. Oxford: Update Software.

15 Anthonisen NR, Connett JE, Kiley JP, Altose MD, *et al.* Effects of smoking intervention and the use of an inhaled anticholinergic bronchodilator on the rate of decline of FEV1. The Lung Health Study. *JAMA* 1994; **272**: 1497–505.

16 Hajek P. Withdrawal-oriented therapy for smokers. *Br J Addict* 1989; **84**: 591–8.

17 West R, Edwards M, Hajek P. A randomised controlled trial of a 'buddy' system to improve success at giving up smoking in general practice. *Addiction* 1998; **93**: 1007–11.

18 Abbot NC, Stead LF, White AR, Barnes J, Ernst E. *Hypnotherapy for smoking cessation (Cochrane review)*. In: The Cochrane Library, Issue 2, 1999. Oxford: Update Software.

19 White AR, Rampes H. *Acupuncture for smoking cessation (Cochrane Review)*. In: The Cochrane Library, Issue 2, 1999. Oxford: Update Software.

20 Russell M. Nicotine intake and its control over smoking. In: Wonnacott S, Russell M, Stolerman I (eds). *Nicotine psychopharmacology*. Oxford: Oxford University Press, 1990.

21 West R. The nicotine replacement paradox in smoking cessation: how does nicotine gum really work? *Br J Addict* 1992; **87**: 165–7.

22 Silagy C, Mant D, Fowler G, Lancaster T. *Nicotine replacement therapy for smoking cessation (Cochrane Review)*. In: The Cochrane Library, Issue 2, 1999. Oxford: Update Software.

23 Stapleton J, Russell M, Feyerabend C, Wiseman S, *et al.* Dose effects and predictors of outcome in a randomized trial of transdermal nicotine patches in general practice. *Addiction* 1995; **90**: 31–42.

24 Hughes J, Lesmes G, Hatsukami D, Richmond R, *et al.* Are higher doses of nicotine replacement more effective for smoking cessation? *Nicotine Tob Res* 1999; **1**: 169–74.

25 Tonnesen P, Paoletti P, Gustavsson G, Russell M, *et al.* Higher dosage nicotine patches increase one-year smoking cessation rates: results from the European CEASE trial. *Eur Resp J* 1999; **13**: 238–46.

26 Transdermal Nicotine Study Group. Transdermal nicotine for smoking cessation: results of two multicenter controlled trials. *JAMA* 1991; **266**: 3133–8.

27 Dale L, Hurt R, Offord K, Lawson G. High-dose nicotine patch therapy: percentage of replacement and smoking cessation. *JAMA* 1995; **274**: 1353–8.

28 Jorenby DE, Smith SS, Fiore MC, Hurt RD, *et al.* Varying nicotine patch dose and type of smoking cessation counseling. *JAMA* 1995; **274**: 1347–52.

29 Hajek P, Jackson P, Belcher M. Long-term use of nicotine chewing gum. Occurrence, determinants, and effect on weight gain. *JAMA* 1988; **260**: 1593–6.

30 Sutherland G, Stapleton J, Russell MAH, Jarvis M, *et al.* Randomized controlled trial of nasal nicotine spray in smoking cessation. *Lancet* 1992; **340**: 324–9.

31 Stapleton J, Sutherland G, Russell M. How much does relapse after one year erode effectiveness of smoking cessation treatments? Long term follow up of randomised trial of nicotine nasal spray. *Br Med J* 1998; **316**: 830–1.

32 Benowitz N (ed). *Nicotine safety and toxicity*. New York: Oxford University Press, 1998.

33 Shiffman S, Gitchell J, Burton S, Kemper K, Lara E. Public health benefit of over-the-counter nicotine medications. *Tob Control* 1997; **6**: 306–10.

34 Benowitz N, Jacob P, Savanapridi C. Determinants of nicotine intake while chewing nicotine polacrilex gum. *Clin Pharmacol Ther* 1987; **41**: 467–73.

35 Hajek P, West R, Foulds J, Nilsson F, *et al.* Randomised comparative trial of nicotine chewing gum, transdermal patch, nasal spray, and inhaler. *Arch Intern Med* 1999; **159**: 2033–8.

36 Daughton DM, Heatley SA, Prendergast JJ, Causey D, *et al.* Effect of transdermal nicotine delivery as an adjunct to low-intervention smoking cessation therapy. A randomized, placebo-controlled, double-blind study. *Arch Intern Med* 1991; **151**: 749–52.

37 Bohandana A, Nilsson F, Martinet Y. Nicotine inhaler and nicotine patch: a combination therapy for smoking cessation. *Nicotine Tob Res* 1999; **1**: 189.

38 Sutherland G. A placebo-controlled double-blind combination trial of nicotine patch and spray. *Nicotine Tob Res* 1999; **1**: 186–7.

39 Blondal T, Gudmundsson J, Olafsdottir I, Gustavsson G, Westin A. Nicotine nasal spray with nicotine patch for smoking cessation: randomised trial with six year follow-up. *Br Med J* 1999; **318**: 285–8.

40 Henningfield JE. Nicotine medications for smoking cessation. *N Engl J Med* 1995; **333**: 1196–203.

41 Hughes JR. Impact of medications on smoking cessation In: Burns D (ed). *Population impact of smoking cessation interventions.* NCI Monograph (in press).

42 Henningfield JE, Fant RV, Gopalan MA. Non-nicotine medications for smoking cessation. *J Respir Dis* 1998; **19**: S33–42.

43 Hughes JR. Non-nicotine pharmacotherapies for smoking cessation. *J Drug Dev* 1994; **6**: 197–203.

44 Benowitz NL. Treating tobacco addiction – nicotine or no nicotine. *N Engl J Med* 1997; **337**: 1230–1.

45 Hughes JR, Stead LF, Lancaster T. *Anxiolytics and antidepressants for smoking cessation (Cochrane review).* In: The Cochrane Library, Issue 3, 1999. Oxford: Update Software.

46 Hughes JR. Comorbidity and smoking. *Nicotine Tob Res* (in press).

47 Goldstein MG. Bupropion sustained release and smoking cessation. *J Clin Psychiatry* 1998; **59** (Suppl 4): 66–72.

48 Hughes JR, Goldstein MG, Hurt RD, Shiffman S. Recent advances in pharmacotherapy of smoking. *JAMA* 1999; **281**: 72–6.

49 Jorenby DE, Leischow SJ, Nides MA, Rennard SI, *et al.* A controlled trial of sustained-release bupropion, a nicotine patch, or both for smoking cessation. *N Engl J Med* 1999; **340**: 685–91.

50 Hall SM, Reus VI, Munoz RF, Sees KL, *et al.* Nortriptyline and cognitive-behavioral therapy in the treatment of cigarette smoking. *Arch Gen Psychiatry* 1998; **55**: 683–90.

51 Prochazka AV, Weaver MJ, Keller RT, Fryer GE, *et al.* A randomized trial of nortriptyline for smoking cessation. *Arch Intern Med* 1998; **158**: 2035–9.

52 Jacobs MA, Spiker AZ, Norman MM, Wohlberg GW, Knapp PH. Interaction of personality and treatment conditions associated with success in a smoking control program. *Psychosom Med* 1971; **6**: 545–56.

53 Edwards NB, Murphy JK, Downs AD, Ackerman BJ, Rosenthal TL. Doxepin as an adjunct to smoking cessation: a double-blind pilot study. *J Psychiatry* 1988; **146**: 373–6.

54 Fowler JS, Volkow ND, Wang GJ, Pappas N, *et al.* Brain monoamine oxidase A inhibition in cigarette smokers. *Proc Natl Acad Sci USA* 1996; **93**: 14065–9.

55 Fowler JS, Volkow ND, Wang GJ, Pappas N, *et al.* Inhibition of monoamine oxidase B in the brains of smokers. *Nature* 1996; **379**: 733–6.

56 Berlin I, Said S, Spreux-Varoquaux O, Olivares R, *et al.* Monoamine oxidase A and B activities in heavy smokers. *Biol Psychiatry* 1995; **38**: 756–61.

57 Niaura R, Goldstein M, Spring B, Keuthen N, *et al.* Fluoxetine for smoking cessation: a multicenter randomized double blind dose response study. *Ann Behav Med* 1997; **19**: S042.

58 Frederick SL, Hall SM, Reus VI, Sees KL. The effect of venlafaxine on smoking cessation in subjects with and without a history of depression. *Problems of drug dependence, 1996.* NIDA Research Monograph 174. Washington, DC: US Government Printing Office, 1997: 208.

59 Gourlay SG, Benowitz NL. Is clonidine an effective smoking cessation therapy? *Drugs* 1995; **50**: 197–207.

60 Clarke PBS. Nicotinic receptor blockade therapy and smoking cessation. *Br J Addict* 1991; **86**: 501–5.

61 Stolerman IP. Could nicotine antagonists be used in smoking cessation. *Br J Addict* 1986; **81**: 47–53.

62 Rose JE, Levin ED. Concurrent agonist-antagonist administration for the analysis and treatment of drug dependence. *Pharmacol Biochem Behav* 1991; **41**: 219–26.

63 Elan Corporation. Elan issues a further report on development pipeline. Press Release, 14 December 1998.

64 West RJ, Hajek P, McNeill A. Effect of buspirone on cigarette withdrawal symptoms and short-term abstinence rates in a smokers clinic. *Psychopharmacology* 1991; **104**: 91–6.

65 Cinciripini PM, Lapitsky L, Seay S, Wallfisch A, *et al.* A placebo-controlled evaluation of the effects of buspirone on smoking cessation: differences between high- and low-anxiety smokers. *J Clin Psychopharmacol* 1995; **15**: 182–91.

66 Schneider NG, Olmstead RE, Steinberg C, Sloan K, *et al.* Efficacy of buspirone in smoking cessation: a placebo-controlled trial. *Clin Pharmacol Ther* 1996; **60**: 568–75.

67 West R, Hajek P. Randomised controlled trial of ondansetron in smoking cessation. *Psychopharmacology* 1996; **126**: 95–6.

68 Damaj MI, Patrick GS, Creasey KR, Martin BR. Pharmacology of lobeline, a nicotinic AChR ligand. *J Pharmacol Exp Ther* 1997; **282**: 410–9.

69 Davison GC, Rosen RC. Lobeline and reduction of cigarette smoking. *Psychol Rep* 1972; **31**: 443–56.

70 Glover ED, Leischow SJ, Rennard SI, Glover PN, *et al.* A smoking cessation trial with lobeline sulfate: a pilot study. *Am J Health Behav* 1998; **22**: 62–74.

71 Covey LS, Glassman AH, Stetner F. Naltrexone effects on short-term and long-term smoking cessation. *J Addict Dis* 1999; **18**: 31–41.

72 O'Malley SS, Krishman-Sarin S, Meandzija B. *Naltrexone treatment of nicotine dependence: a preliminary study.* Paper presented at the Society of Research on Nicotine and Tobacco, Nashville, TN, 1997.

73 Wong GY, Wolter TD, Croghan GA, Croghan IT, *et al.* Randomized trial of naltrexone for smoking cessation. *Addiction* 1999; **94**: 1227–37.

74 US Department of Health and Human Services. Smoking deterrent drug products for over-the-counter human use: establishment of a monograph. (Docket No. 81N-0027). *Fed Register* 1982;**47**: 490–500.

75 Hymowitz N, Feuerman M, Hollander M, Frances RJ. Smoking deterrence using silver acetate. *Hosp Community Psychiatry* 1993; **44**: 113–7.

76 Westman EC, Behm FM, Rose JE. Airway sensory replacement as a treatment for smoking cessation. *Drug Dev Res* 1996; **38**: 257–62.

77 Levin ED, Behm F, Carnahan E, LeClair R, *et al.* Clinical trials using ascorbic acid aerosol to aid smoking cessation. *Drug Alcohol Depend* 1993; **33**: 211–23.

78 Behm FM, Schur C, Levin ED, Tashkin DP, Rose JE. Clinical evaluation of a citric acid inhaler for smoking cessation. *Drug Alcohol Depend* 1993; **31**: 131–8.

79 Westman EC, Behm FM, Rose JE. Airway sensory replacement combined with nicotine replacement for smoking cessation. *Chest* 1995; **107**: 1358–64.

80 Raw M, McNeill A, West R. Smoking cessation guidelines for health professionals. A guide to effective smoking cessation interventions for the health care system. *Thorax* 1998; 53(Suppl 5, Part 1).

81   Parrott S, Godfrey C, Raw M, West R, McNeill A. Guidance for commissioners on the cost-effectiveness of smoking cessation interventions. *Thorax* 1998; **53**(Suppl 5, Part 2).

82   Health Education Authority. *Health in England*. London: HEA, 1995.

83   Bolling K, Owen L. *Smoking and pregnancy. A survey of knowledge, attitudes and behaviour*. London: Health Education Authority, 1997.

84   Godfrey C, Raw M, Burrows S. *An estimate of national expenditure on smoking cessation in the UK.* York: Centre for Health Economics, University of York, 1998.

85   Owen L, Scott P. Barriers to good practice in smoking cessation work among pregnant women. *J Inst Health Educ* 1995; **33**: 110–2.

86   Department of Health. *Smoking kills. A White Paper on tobacco*. London: The Stationery Office, 1998.

87   Cheung AM, Tsevat J. Economic evaluations of smoking interventions. *Prev Med* 1997; **26**: 271–3.

88   Warner KE. Cost-effectiveness of smoking cessation therapies: interpretation of the evidence and implications for coverage. *Pharmacoeconomics* 1997; **11**: 538–49.

89   Buck D, Godfrey C. *Helping smokers give up: guidance for purchasers on cost-effectiveness.* London: Health Education Authority, 1994.

90   Stapleton JA, Lowin A, Russell MAH. Prescription of transdermal nicotine patches for smoking cessation in general practice: evaluation of cost effectiveness. *Lancet* 1999; **354**: 210–5.

91   Curry SJ, Grothaus LC, McAfee T, Pabiniak C. Use and cost effectiveness of smoking-cessation services under four insurance plans in a health maintenance organization. *N Engl J Med* 1998; **339**: 673–9.

92   Lightwood JM, Glantz SA. Short-term economic and health benefits of smoking cessation: myocardial infarction and stroke. *Circulation* 1997; **94**: 1089–96.

93   Krumholz HM, Cohen BJ, Tsevat J, Pasternak RC, Weinstein MC. Cost effectiveness of a smoking cessation program after myocardial infarction. *J Am Coll Cardiol* 1993; **22**: 1697–702.

94   Maynard A. Developing the health care market. *Econ J* 1991; **101**: 1277–86.

95   Tengs TO, Adams ME, Pliskin JS, Safran DG, *et al.* Five-hundred life saving interventions and their cost-effectiveness. *Risk Anal* 1995; **15**: 369–89.

8 | Regulatory approaches to tobacco products in Britain

8.1 The evolution of British laws and voluntary agreements

Cigarettes are exempt from most forms of consumer protection legislation. This anachronistic situation has arisen, first, because cigarettes were already on the market when consumer protection laws were being developed and, secondly, because the extent of the harm caused by cigarettes to the consumer is such that they do not, and without radical changes probably cannot, meet the safety requirements imposed on other products.

On occasions when the tobacco industry has faced the threat of product legislation, its response has been to make voluntary concessions. Over the years, these have resulted in a complex set of regulations known as 'voluntary agreements', negotiated in private between representative bodies of the industry and government officials. The choice of voluntary agreements with, rather than legislation on, the tobacco industry also partly reflects a British tradition to seek solutions by consensus. However, whilst the voluntary system gave an impression of regulation and an air of respectability to the industry, it was fatally flawed both because it was industry led and because the partial controls involved were easily evaded. The agreements themselves were difficult to monitor and there were no real penalties for violations. For many years, the only exceptions to the voluntary agreement approach were in the areas of fiscal policy and in the supply of tobacco to children (which has been restricted by law since 1908).

The entry of the UK into the European Community (now the European Union (EU)) largely brought about an end to the voluntary system of control. The introduction of the single market in 1992 meant that laws affecting trade had to be harmonised, with the standard chosen being that providing the highest level of health protection. Hence, in situations in which the EU had legislated, the old voluntary agreement in that area would be superseded, requiring the UK to introduce legislation. In December 1998, the government produced the first ever White Paper on Tobacco, *Smoking kills*,[1] which set out a wide range of measures to control tobacco, including support for the recent European Directive (98/43/EC) to ban advertising and sponsorship of tobacco products in the EU. The regulation of tobacco products in Britain is therefore currently undergoing a period of change, but throughout this century cigarettes have enjoyed a relative freedom from legislative control.

8.2 The regulation of nicotine, tar and additives in cigarettes

Filters, low tar cigarettes and the Federal Trade Commission test

Ever since health concerns regarding smoking were first raised in the 1950s, tobacco product manufacturers have responded by trying to make their products appear to be less dangerous.[2,3] The first development in this respect was the introduction of filter tips in the mid-1950s, and subsequently ventilated filters in the 1970s. The second major development was the introduction of 'low tar' brands of cigarettes, a development which has continued to the present day, but which has failed to deliver substantially less toxicity.[4,5] As discussed in earlier sections of this report, such initiatives may well have helped to keep people smoking by providing 'health reassurance'.

In response to health claims made by the tobacco industry for lower tar cigarettes, in 1967 the US Federal Trade Commission (FTC) established a standardised protocol for assessing tar and nicotine yields. Now commonly known as the FTC test, it was adopted by the International Standards Organization (ISO) and later broadened to include measurement of carbon monoxide (CO) yield. Routine monitoring of tar and nicotine yields of cigarette brands in Britain began in 1972, and in 1975 the industry agreed to display tar bands (high, medium, low) on packets.

Tobacco substitutes

Following the introduction of filter and low tar cigarettes, the tobacco industry considered the value of introducing synthetic tobacco substitutes into cigarettes. A proportion of the natural tobacco was replaced with material of lower biological activity, offering the prospect of reducing the toxicity of the cigarettes. Having abided by guidelines for the testing of substitutes,[6] the tobacco industry marketed the first two major tobacco substitutes: Cytrel 361 and 'New Smoking Material' (NSM 14) in 1977. These were a commercial failure, and were eventually withdrawn. A large-scale epidemiological study to examine the long-term impact on health of these products was also abandoned as it failed to enlist an adequate sample of substitute cigarette smokers.

New voluntary agreements on tar reductions

Following the failure of tobacco substitutes in the 1970s, new voluntary agreements on product modification were drawn up in 1980, mainly with a view to achieving reductions in the tar yields of British cigarettes. The 1980 agreement adopted a target of 15 mg sales-weighted average tar yield (SWAT) per cigarette to be achieved by the end of 1983, and the 1984 voluntary agreement proposed a reduction of SWAT to about 13 mg per cigarette by the end of 1987. No limits for nicotine were set. It was suggested that CO yields should be reduced to the lowest possible levels.[7] In the 1970s, it was also suggested that cigarette smoking could be made less hazardous by reducing tar and other toxins relative to nicotine.[8] Although this was never explicitly adopted as a public health strategy, the UK government recognised its potential advantages in light of the compensatory nature of people's smoking. It therefore tolerated reductions in the average SWAT/ nicotine (T/N) ratio, which declined between 1972 and 1987.[9]

As stated above, following the UK's entry to the EU the voluntary agreements on reductions in tar yields were superseded by EU regulations. These have progressively reduced the tar yield of manufactured cigarettes, the most recent regulation stipulating an upper limit of 12 mg per cigarette by 31 December 1997. There is now a clear downward trend in nicotine yields in the UK – and indeed across the EU – in line with the enforced reduction in tar yields.

A separate EU directive on labelling required FTC/ISO tar and nicotine yields to be printed on the side of cigarette packets. It is not clear whether smokers notice this information or whether the information has any meaning to them,[10] but this change led to the end of the tar band system in the UK. Tobacco companies began to use descriptive terms such as 'light' and 'mild' for lower delivery brands. Low tar cigarettes (defined as those yielding less than 10 mg tar) began to take an increasing market share.

However, FTC/ISO measures of tar and nicotine yields from cigarettes smoked by machines bear little relation to the actual amounts obtained from cigarettes by human smokers because, as argued in Chapter 6 and explicitly acknowledged by the FTC,[11] they do not allow for compensation in smoking behaviour. It was also argued in Chapter 6 that there is no strong evidence of any reduction in actual exposure of smokers to tar since the 1960s. However, many consumers still believe that low tar cigarettes are safer.[12]

The introduction of additives

Prior to 1970, the use of additives in UK cigarettes was largely proscribed. The Finance Act (1970) enabled tobacco duty to be charged on additives and substitutes used in the manufacture of smoking products;[6] this, in turn, enabled the Commissioners of Customs and Excise to relax the restrictions on the ingredients that could be used in tobacco manufacture. Statutory control over the materials used in the manufacture of tobacco products finally ceased with a radical revision of the tax system in 1978.

A voluntary agreement was concluded in 1977 in which the tobacco industry agreed not to introduce new products containing additives other than those found acceptable to the then scientific committee, the Independent Scientific Committee on Smoking and Health (ISCSH). The government stated, however, that it would amend the Medicines Act (1968) to control the use of tobacco substitutes and additives in smoking products in the UK if the need arose at any time.[6]

In 1978, a list of permitted additives was drawn up.[6] The ISCSH recognised the potential value of using flavouring additives to improve the acceptability of low tar cigarettes, but stated that no additional 'dependence inducing' compounds could be incorporated into tobacco. It also recognised that additives could be used in the manufacture of cigarettes for technological reasons, provided they were safe, for example, to:

* prevent the fall of ash
* control the rate of burning, and
* inhibit the formation of mould.

At that time, the term 'additive' referred to any substance added in the course of manufacture of a smoking product to alter the smoking quality, appearance or any other characteristic of that product. It excluded additives in filters, cigarette papers, filter wrappers, tips and overwrappers. The definition was extended in 1983 to include assessment of all parts intended to be burnt, so cigarette papers were then encompassed within the voluntary agreements.[7]

A further voluntary agreement for the approval of new additives to tobacco was agreed in 1997, replacing the 1984 agreement. While the advisory scientific committee of the Department of Health (DoH) (the Scientific Committee on Tobacco and Health) expressed a clear reservation about the possibility that additives could prolong the use of cigarettes by making them more palatable, the committee recommended only that the use of additives in tobacco products be closely monitored.[12] The new voluntary agreement required tobacco manufacturers to provide toxicological and other data for any new additives that they wished to add to the approved list, but such information was not required for the 600 existing approved additives.

There has been no systematic evaluation of the overall public health impact of additives. Even if lacking direct toxicity, they may well add to the burden of harm caused by tobacco use by making cigarettes more palatable, attractive or addictive. Little is known about the potential harmful effects of many additives when they are burnt with tobacco or in conjunction with other additives. Even for new additives, no criteria are set out for measuring their public health impact.[13] This is one of several highlighted areas of concern regarding the influence of additives on smoking behaviour.[13]

A further weakness of the voluntary agreement on additives is that it can be circumvented by securing approval for the additive in a different EU member state. Directive 83/189/EEC requires the DoH to 'raise no objection' to the use of an additive permitted in any other EU member state provided that certain specified information is supplied.[13] However, recent cases in other areas indicate that there may be provision in the law for a country to refuse an additive that is allowed in another country.

Health warnings

In 1971, a voluntary agreement was drawn up between the industry and government which specified that cigarette packs for sale in the UK should carry a health warning. Subsequent agreements have extended and strengthened these warnings.

In 1991, the government announced a series of new, larger health warnings for tobacco packaging in line with EU requirements (Tobacco Products Labelling (Safety) Regulations 1991). By the new regulation, cigarette packs were legally required to carry two health warnings which should cover 6% of the relevant face of the pack (the minimum requirement under the terms of the Directive was 4%). All tobacco products were to carry the general warning '*Tobacco seriously damages health*', and the additional warning was to be chosen from a set list of 15. The UK government rotates six of the 15 warnings, but none of these six includes the warnings related to addiction.

Although intended to discourage smoking, there is evidence that smokers pay little heed to health warnings and do not remember them. Paradoxically, health warnings may have given an advantage to the tobacco manufacturers because they can be used as evidence that consumers were warned, and hence as a defence in litigation.

Other nicotine containing products

Oral snuff. In February 1998, the UK government announced a ban on the supply and sale of oral snuff under the 1987 Consumer Protection Act. The regulations which prohibited the manufacture or storage (for supply to other countries) of oral snuff were challenged by the tobacco industry, but formed the basis for the later ban (89/622/EEC) on oral snuff throughout the EU.

Nicotine treatment products. Nicotine treatment products are regulated by the Medicines Control Agency, and are therefore subject to the rigorous regulatory standards and assessment governing pharmaceutical products in the UK (see Chapter 4, Table 4.2 for summary). Any additives must be disclosed by weight in each specific product, and their use justified. Most forms of nicotine replacement therapy are licensed for sale in only a restricted range of outlets, in the case of nicotine nasal spray only on prescription. This paradox arises from the fact that tobacco products are exempt from the regulations that apply to pharmaceutical nicotine products.

Conclusions

Nicotine regulation in the UK has so far mainly served the interests of the tobacco industry rather than those of public health. Low tar cigarettes, additives and health warnings have all been turned to the industry's advantage, despite the best intentions of government. The voluntary agreement approach has been discredited. European directives were a step forward in that they were mandatory, but they were built on measures such as the FTC/ISO test, which is now known to be crucially flawed as a guide to cigarette toxicity.[11] A new regulatory framework for nicotine is urgently needed. The following sections discuss the challenges facing future regulators.

8.3 Anomalies in current nicotine regulation

While nicotine itself cannot be completely exonerated from causing adverse health effects, it is clear that its direct contribution to tobacco related harm is relatively minor. It is not nicotine itself, but the delivery of nicotine in combination with a multitude of combustion products that causes most of the deaths associated with tobacco use.[14] Tobacco products, particularly cigarettes, are an exceedingly 'dirty' delivery system for nicotine.[15] The existing regulatory structures give huge market-place advantages to tobacco products, effectively creating a 'nicotine maintenance monopoly'. Those who want or need nicotine on an ongoing basis have little choice but to obtain it from cigarettes – the contaminated delivery system, with an attendant 50% risk of premature death.[16]

Nicotine delivery products which do not qualify as tobacco products and are not pharmaceutical products are also at great risk of being banned pursuant to consumer protection and poisons laws (see below). The result of this regulatory environment is that tobacco products are given an overwhelming market-place advantage over the other nicotine delivery systems.[4]

There is an overwhelming case for ending this regulatory advantage. The present 'upside down' regulation framework for nicotine products should be adjusted to promote health, without relaxing the standards for medicines. Adjusting regulations to allow a wider range of more consumer-acceptable smoking cessation and smoking reduction products, and distinguishing between smoking cessation and nicotine cessation, may stimulate the necessary economic incentives to market products that would significantly reduce tobacco's death toll.

8.4 Novel nicotine delivery devices from tobacco companies

In addition to filters, low tar cigarettes and tobacco substitutes, less toxic cigarette-like nicotine delivery devices have been developed and test marketed by the tobacco industry.[17,18] In the late 1980s, RJ Reynolds launched a new era of novel devices with a product called 'Premier',[19] which was the size and shape of a cigarette. Premier used heat created by burning a piece of carbon at its distal end to vaporise nicotine and glycerine absorbed on to alumina pellets in an aluminium tube adjacent to the heat source. The rest of the device acted as a cooling chamber to permit the formation of an aerosol. Premier delivered doses of nicotine comparable to those of a cigarette with little other toxic material except large doses of CO. Its test marketing was cut short, however, in the face of poor sales, complaints about its taste and difficulty lighting, the threat of Food and Drug Administration (FDA) regulation as a drug, and reports that it could be used to smoke crack.[20]

Using much the same technology, RJ Reynolds has developed another novel product, marketed in the US as 'Eclipse'.[21,22] In Eclipse, nicotine and glycerine are vaporised from an aluminium-lined chamber filled with a shredded paper derived from tobacco and mixed about 1:1 with glycerine. The proximal part of the device again acts as a cooling chamber. While toxin delivery is higher from Eclipse than from Premier, it is lower than that from most, but not all, marketed cigarettes.[23] Once again, nicotine delivery is comparable to that of a cigarette, and CO delivery is very high. In addition, glass fibres from the insulator around the fuel element can become dislodged, and some are loosely adherent to the mouthpiece.[24] A short-term study found evidence of reduced inflammation in the tracheo-bronchial tree among subjects smoking Eclipse compared to their usual brand of cigarette.[25]

Eclipse has been test marketed for three years. Many smokers tried the product, but few stayed with it. Claims have been made in advertisements for reduced environmental tobacco smoke, and that the product produces less smoke smell and stain than conventional cigarettes whilst still producing satisfaction.

Philip Morris has made similar claims for its novel product, 'Accord'. Accord consists of two parts: a specially constructed cigarette, and a unique lighter into which the cigarette must be inserted. Taking a puff on Accord activates one of eight radially arrayed heating elements in the lighter that scorches a narrow segment of the cigarette. These eight heating elements are activated sequentially. The aerosol from Accord has a chemical profile similar to that of Eclipse, except that it has a substantially lower yield of CO.[26,27] As with Eclipse, sales seem to have been sluggish in test markets.

Were these products to be marketed with promises, direct or implied, of substantially reduced risk to health, sales might be expected to rise substantially. Whether these products remain niche curiosities or become major players in the market may depend on the claims that their makers assert for them. However, these novel nicotine delivery devices raise a number of regulatory concerns.[5,28] One important question is whether the products have less toxicity than conventional products. A review of product design, chemical analyses and various toxicological studies could address the most basic aspects of this question. More difficult to answer, though, is the nature and extent of any long-term problems. Chronic toxicity may not be revealed in short-term tests, and products such as these could conceivably – as with 'light' cigarettes – make the overall problem worse by leading to a larger total market than would otherwise be the case.[4,5] It will be especially important to regulate products such as these (as well as other tobacco products) in parallel with the regulation of medicinal forms of nicotine because of the interrelated nature of the overall market for nicotine delivery devices.[4]

An approach to this dilemma might be to grant preliminary approval for marketing of products such as these, based on a range of information and assurances from the manufacturers.[5] Post-marketing studies by the manufacturer would be required so that adverse effects on public health could be detected and promptly corrected.

Similar regulatory issues are posed by the question of how to deal with a new tobacco curing process that substantially reduces nitrosamines.[29] While the tobacco-specific nitrosamines are potent carcinogens, and clearly should be removed from tobacco if at all possible, publicity or marketing claims about this could raise false public expectations about enhanced safety.

The development of Eclipse, Accord and low nitrosamine tobacco indicates that potentially useful technological innovation for tobacco products is possible. This was also demonstrated by a report by the Imperial Cancer Research Fund and Action on Smoking and Health, showing that the tobacco industry had investigated and patented many technologies that would reduce substances in cigarette smoke that cause cancer, heart disease and emphysema.[30]

The challenge faced by public health authorities is how to both encourage and channel innovation so that it leads to real public health benefits by reducing the illness and death caused by tobacco products. Left to their own devices or, worse still, if only loosely regulated, tobacco product manufacturers have shown in the past that they will put profits ahead of public health in their use of innovations.

8.5 Tobacco regulation in the USA

In early 1994, the US FDA launched an investigation of cigarettes and smokeless tobacco products,[31] with two sequential aims:

1   To determine if these products came within the jurisdiction of the FDA.
2   To regulate these products in ways that protected the public health.

Since the 1950s, the agency had occasionally regulated individual tobacco products on a case-by-case basis, but had never asserted jurisdiction over a whole class of products. The 18-month investigation led to an assertion of jurisdiction and proposed regulations.[32] The jurisdictional determination was based on converging evidence that established that cigarettes and smokeless tobacco produced pharmacological effects, including addiction, in consumers, and that the manufacturers of these products intended these effects.

Industry manipulation of tobacco

Much of the evidence on which the agency relied was newly revealed as a result of litigation and indicated that cigarettes are highly refined, carefully engineered products. A substantial part of these data showed that the modern manufactured cigarette depends on numerous manipulations to produce a consistent product that reliably delivers doses of nicotine in the desired range.[18]

In testimony before Congress in March 1994, FDA Commissioner David Kessler compared the tobacco industry to the pharmaceutical industry in its emphasis on the bioavailability of nicotine and its exploration of nicotine analogues.[33] He reported that cigarettes in the US market have contained gradually increasing nicotine-to-tar ratios, a phenomenon which requires substantial control over numerous aspects of cigarette design and construction. The FDA also found that the nicotine contents of the tobacco in three sub-brands of Merit (Philip Morris) were inversely proportional to the machine-measured 'tar' delivery of the sub-brands.[32,33]

The use of nicotine-rich tobaccos in cigarettes designed to have lower 'tar' yields has been a regular feature in the industry for decades.[34] William Farone, a former director of applied research at Philip Morris, reflecting on his years in the business, noted:

> Product developers and blend and leaf specialists were responsible for manipulating and controlling the design and production of cigarettes in order to satisfy the consumer's need for nicotine in lower yield products.
>
> Blend changes were an especially important tool used to ensure desired nicotine levels.[34]

Dr Kessler also referred to the development of 'Y-1' tobacco by the Brown & Williamson Tobacco Company (a subsidiary of British American Tobacco).[34,35] Using advanced biotechnology, the company engineered a strain of tobacco that produces leaf with a substantially higher nicotine content.

The Food and Drug Administration tobacco rule

The FDA made an initial determination that cigarettes and smokeless tobacco products fit the legal definitions of both drugs and devices.

The agency decided to regulate them as devices since the rules for drugs would have required these products to be banned as unsafe. Following extensive public consultation, the FDA issued a final rule in August 1996.[34]

Since then, the matter has been in court. A district judge upheld the agency's jurisdiction, but an appeals court ruled in favour of the industry. The issue is currently before the Supreme Court, with a decision expected by June 2000. If the Supreme Court decides that the FDA has jurisdiction, the agency can proceed with regulation of tobacco products. If it rules against the FDA, the only way the agency can regulate tobacco products will be by the enactment of new legislation by Congress.

The FDA's initial regulations focused on preventing the uptake of tobacco use by young people. Their main features were modest limits on advertising, and a series of measures establishing a federal age of sale (18 years) and related enforcement provisions. However, under current law, the agency could regulate virtually every aspect of the production and marketing of tobacco products.[5,36] It could establish minimum standards for the composition of these products, and require the disclosure of information to consumers about them. A key area would be the regulation of health claims for tobacco products. The false expectations generated by filtered cigarettes and 'light' cigarettes[4,5] have made it clear that tobacco product manufacturers must be accountable both for the claims they make and also for the effects of these claims on the consumption of their products.

The FDA also regulates medicinal forms of nicotine. While tobacco products may never be subject to the stringent rules that are applied to medicines, it is possible for the agency to 'coregulate' nicotine delivery devices from tobacco companies, on the one hand, and nicotine delivery devices from pharmaceutical companies, on the other, to achieve optimal public health benefit.[28]

The Federal Trade Commission test

In September 1997, the US FTC solicited comments on proposed revisions to the FTC test, acknowledging the limitations of the existing test method. Their proposed revisions included:

- measuring tar, nicotine and CO yields obtained under two different smoking conditions
- public education to make smokers aware that any benefits of switching to lower yield cigarettes are small compared with quitting, and

- a re-examination of the system at least every five years to evaluate whether the protocol is maintaining its utility to the smoker.

These proposals were strongly criticised by US scientists as not being sufficient either to provide consumers with accurate information or to give appropriate parameters for ensuring future reductions in harm. The FTC has now withdrawn the proposed new methodology and has suggested a rethink lasting 18 months.

8.6 Future nicotine regulation in the UK

The evidence reviewed in the preceding sections demonstrates that cigarettes are highly engineered drug delivery devices which remain largely unregulated. Steps have been taken in the US to bring cigarettes under the auspices of the FDA, enabling all nicotine containing products to be regulated by one body. The question now arises as to whether there is scope for similar regulation in the UK.

Regulation at a national level

A common regulatory framework for nicotine products could be introduced in the UK by an extension to pharmaceutical regulation, the Medicines Act (1968), to include all nicotine containing products, or by introducing new legislation to develop a new nicotine regulatory authority.

In the mid-1970s, the then Health Minister, Dr David Owen, made an attempt to bring cigarettes under the Medicines Act. Section 105 of this Act empowered the Minister to include by order in the Act any substance which, while not itself a medicine, nevertheless:

> if used without proper safeguards, is capable of causing danger to the health of the community.

By tabling an order decreeing tobacco to be such a substance, the Medicines Act could be used to control cigarettes. These plans were dropped following ministerial reshuffles.

Regulation at a European level

In February 1999, Commissioner Flynn announced that the European Commission was considering legislation concerning tar, nicotine and CO yields, as well as new provisions to regulate additives and the labelling of tobacco products. This could also be a useful route to introduce an improved regulatory framework for nicotine products throughout the EU.

The scope of future nicotine regulation

Some broad conclusions can be drawn about the direction of future nicotine regulation:

- It is clear that the FTC test should no longer be used as the basis for product regulation in the long term. New protocols for measuring cigarette emissions should take account of recommendations emanating from the 18-month review now underway in the US.

- The concept of 'tar' as an undifferentiated whole is misleading, and should not be used for regulatory purposes. Tar has markedly different compositions between products which are likely to cause different degrees of harm. For example, most of the tar fraction from Eclipse is actually glycerine.

- The branding of low tar cigarettes, using implied health claim terms such as 'light' or 'mild' is completely unjustified.

- Tar and nicotine yields based on the existing FTC measurements provide no meaningful information for consumers and should be removed from the pack.

- The use of additives in cigarettes was justified on the grounds that they might facilitate the acceptance of low-yield cigarettes. As low tar cigarettes have now been discredited, it is difficult to justify the use of additives on this basis.

- Before meaningful regulation can be introduced, regulators need a detailed characterisation of cigarette products. A precedent for this has been set by the government of British Columbia.[37]

- The separate regulatory systems for treatments and tobacco products have thus far given great advantage to tobacco. Cigarettes and other tobacco products should, in fact, be subject to the same safety requirements as any other drug delivery device.

- For the future a common nicotine framework should be established to prioritise adjustment of perverse regulatory imbalances that favour the dirtiest nicotine delivery over cleaner forms. It could then enable common approaches to be established to test whether new product developments are in the public interest.

As nicotine contributes to harm from tobacco mainly by creating and sustaining addiction, but does not appear to make a major direct contribution to the adverse health effects of tobacco, there is some debate about the appropriate direction of future nicotine regulation. On the basis that nicotine addiction is the underlying cause of tobacco-related harm, some have argued that permitted nicotine delivery from cigarettes, as determined by absolute nicotine bioavailability rather than by machine-smoking yields, should be progressively reduced to a non-addictive threshold.[38] Others have contended that, since the harm caused by smoking is mainly attributable to the contaminated delivery system (the cigarette) rather than to the drug itself (nicotine), regulation should seek to limit users' exposure to cigarette toxins (tar and gas phase components) while leaving nicotine itself largely unregulated.[39] The future regulation of nicotine products needs to take account of these views, while seeking the most effective regulatory framework to improve the public health in Britain.

Conclusions

The measures so far introduced by governments to control the cigarette product have generally failed to deliver significant public health benefits. Technological developments in nicotine delivery systems point to an emerging convergence between products aimed at satisfying consumers' addiction and pharmaceutical cessation aids. In principle, these products should all, therefore, be subject to the same regulations. A coordinated nicotine regulatory framework has the potential markedly to reduce tobacco-related harms, and should be an urgent priority for the UK government.

8.7 Economic effects of nicotine regulation

One of the major arguments that has always been raised about any regulation of the tobacco industry has been the economic impact of such regulation. Particular concern has been raised about the potential loss of jobs and government revenue. What is the evidence of such economic impact, and can the potential impact of nicotine regulation be predicted?

Employment

The tobacco industry is itself only a small employer in the UK, with less than 12,000 employees in 1992. There have been large job losses in the industry, but this has been due to productivity changes through increased mechanisation, with any fall in domestic consumption being offset by increased exports. Although reducing sales in the UK as a result of any new regulation may result in some job losses in retail and other support industries, reduced tobacco consumption has other economic consequences. Those who stop smoking have more income to spend, and these changes in consumption habits also impact on employment. A study in the UK[40] estimated that a 40% fall in tobacco consumption would result in a *gain* of 115,000 full-time equivalent jobs – the result of those quitting smoking spending more on travel, recreation, meals out and other services. These items of expenditure create more jobs than those lost from tobacco related industry.

Similar conclusions have been reached in other countries. A recent report from the World Bank[41] concluded that a global fall in tobacco consumption would mean that 'more jobs are likely to be created than are lost'.

Tax revenue

Part of the adjustment to major declines in tobacco consumption will be some fall in tobacco tax revenue. While tax receipts have been boosted by increased tax rates in the last few years, over a longer time frame the real value of tobacco taxation has fallen. The importance of tobacco revenue has also fallen dramatically over the longer term as new sources such as Value Added Tax have been introduced. Thus, tobacco tax accounted for 29% of all expenditure taxes in 1960 and 12% of all government current receipts, but by 1995 the corresponding figures were 14% and 3.6%, respectively. Clearly, governments like to preserve their sources of revenue, but can recoup any shortfall from one source. Any reduction in revenue from the effects of regulation is likely to be gradual rather than sudden. Adjustment can therefore be planned by the government; as a result, long-term reduction in the real revenue yield is unlikely to have any major economic consequences.

Regulatory changes involve economic costs as well as benefits. In the main, the costs of changes in manufacture are borne by the industry, but additional resources may be required from government to enforce new regulations. Industry may well seek to pass on any cost of change through increased prices, which has the additional benefit of bringing further reductions in consumption and consequent increases in health benefits.

One potential cost may, however, be to increase incentives to smuggle goods from less regulated parts of the world. Analysis of the patterns of smuggling suggest that the issue is complex. Joossens and Raw[42] found that within Europe it was the countries with the lowest rates of cigarette prices, rather than the highest, where smuggling was highest. It would seem unlikely therefore that changes in regulation would have a significant impact on smuggling. Also, the conclusion from most research is that the best approach is to tackle the crime of smuggling directly and at an international level.[41]

Other effects of smoking reduction

Reductions in smoking may have a number of other consequences. Clearly, people who stop smoking live longer, and are able to contribute to society for a longer period as well as being alive to draw on pensions and make use of health services. It is sometimes argued that there is some sort of economic loss if smokers choose to give up and benefit from a longer life. This argument could be extended to all policies which help to save lives, especially young lives. Clearly, balancing items such as contributions to tax and pensions and individuals' expenditure is one of fairness. The economic argument is around the use of scarce resources, and needs to take account of the value of the life both to the smoker and society. Prabhat and Chaloupka set out the value for money of tobacco control measures.[41] Their figures do not specifically consider changes in regulation, but indicate that tobacco control measures are a highly cost-effective means of improving health in low-, middle- and high-income countries.

Conclusions

While economic impacts have been used to argue against tobacco control measures, independent research suggests that the overall economic impact of reductions in tobacco consumption is likely to be small, even for cigarette producing countries such as the UK. More specifically, the result of people stopping smoking in the UK will be lower demands on the NHS for treatment for smoking related diseases, and more UK employment. The government will get good value for money from additional policies to regulate nicotine, saving many lives at a small cost.

References

1 Department of Health. *Smoking kills. A White Paper on tobacco.* London: The Stationery Office, 1998.
2 Slade JD. The tobacco epidemic: lessons from history. *J Psychoactive Drug* 1989; **21**: 281–91.
3 Kluger R. *Ashes to ashes: America's hundred-year cigarette war, the public health, and the unabashed triumph of Philip Morris.* New York: Knopf, 1996.
4 Warner KE, Slade J, Sweanor DT. The emerging market for long-term nicotine maintenance. *JAMA* 1997; **278**: 1087–92.
5 Slade J, Henningfield JE. Tobacco product regulation: context and issues. *Food Drug Law J* 1998; **53**(Suppl): 43–74.
6 Independent Scientific Committee on Smoking and Health. *Second report.* London: HMSO, 1979.
7 Independent Scientific Committee on Smoking and Health. *Third report.* London: HMSO, 1983.

8   Russell MAH. Low-tar medium-nicotine cigarettes: a new approach to safer smoking. *Br Med J* 1976; **1**: 1430–3.

9   Jarvis M, Russell MAH. Data note - 4. Sales-weighted tar, nicotine and carbon monoxide yields of U.K. cigarettes: 1985. *Br J Addict* 1986; **81**: 579–81.

10 Evans N, Joossens L. *Consumers and the changing cigarette*. London: Health Education Authority, 1999.

11 Federal Trade Commission. 'Tar', nicotine and carbon monoxide of the smoke of 1252 varieties of domestic cigarettes for the year 1997. Issued 1999. www.ftc.gov/os/1999/9909/1997tnrpt.pdf.

12 Department of Health, Department of Health and Social Services, Northern Ireland, The Scottish Office Department of Health, Welsh Office. *The Report of the Scientific Committee on Tobacco and Health*. London: The Stationery Office, 1988.

13 Bates C, Connolly GN, Jarvis M. *Tobacco Additives. Cigarette engineering and nicotine addiction*. London: Action on Smoking and Health, Imperial Cancer Research Fund, 1999.

14 Benowitz N. Pharmacology of nicotine: addiction and therapeutics. *Annu Rev Pharmacol Toxicol* 1996; **36**: 597–613.

15 US Department of Health and Human Services. *Health benefits of smoking cessation. Report of the Surgeon-General*. DHSS Publ No. CDC 90–84 16. Rockville, MD: USDHHS, 1990.

16 Peto R, Lopez AD, Boreham J, Thun M, Heath C. *Mortality from smoking in developed countries*. Oxford: Oxford University Press, 1994.

17 Glantz SA, Slade J, Bero LA, Hanauer P, Barnes DE. *The cigarette papers*. Berkeley, CA: University of California Press, 1996.

18 Slade J. Nicotine delivery devices. In: Orleans CT, Slade J (eds). *Nicotine addiction. Principles and management*. New York: Oxford University Press, 1993: 2–23.

19 RJ Reynolds Tobacco Company. *Chemical and biological studies on new cigarette prototypes that heat instead of burn tobacco*. Winston-Salem, NC: RJ Reynolds Tobacco Co, 1988.

20 Cone EJ, Henningfield JE. Premier 'smokeless cigarettes' can be used to deliver crack. *JAMA* 1989; **261**: 41.

21 Pauly JL, Streck RJ, Cummings KM. US patents shed light on Eclipse and future cigarettes. *Tob Control* 1995; **4**: 261–5.

22 Slade J. Innovative nicotine delivery devices from tobacco companies. In: Ferrence R, Slade J, Pope M, Room R (eds). *Nicotine and public health*. Washington: American Public Health Association (in press).

23 Borgerding MF, Bodnar JA, Chung HL, Morrison CC, *et al. Investigation of a new cigarette which primarily heats tobacco using an alternative puffing regimen*. Presented at the 50th Tobacco Chemists' Research Conference. Richmond, VA, 23 October 1996.

24 Pauly JL, Lee HJ, Hurley EL, Cummings KM, *et al.* Glass fiber contamination of cigarette filters: an additional health risk to the smoker? *Cancer Epidemiol Biomarkers Prev* 1998; **7**: 967–79.

25 Rennard SI, Umino T, Millatmal T, Daughton DM, *et al. Switching to Eclipse is associated with reduced inflammation in the lower respiratory tract of heavy smokers*. Presented at the Society for Research on Nicotine and Tobacco. San Diego, CA, 5–7 March 1999.

26 Terpstra PM, Reininghaus W, Solana RP. *Evaluation of an electrically heated cigarette*. Presented at the Society of Toxicology. Seattle, WA, 5 March 1998.

27 Buchhalter AR, Eissenberg T. *Reduced CO boost produced in humans smoking a novel smoking system*. Presented at the Society for Research on Nicotine and Tobacco. San Diego, CA, 5–7 March 1999.

28   Henningfield JE, Slade J. Tobacco-dependence medications: public health and regulatory issues. *Food Drug Law J* 1998; **53**(Suppl): 75–114.

29   Hwang SL. Latest move to make a safer smoke uses special tobacco. *Wall Street J*, 29 April 1999: B1.

30   Bates C, Jarvis M. *The safer cigarette. What the tobacco industry could do … and why it hasn't done it*. London: Action on Smoking and Health, Imperial Cancer Research Fund, 1999.

31   Kessler DA. *Letter to Scott Ballin, Coalition on Smoking OR Health*. Rockville, MD: US Food and Drug Administration, 25 February 1994.

32   Food and Drug Administration. Regulations restricting the sale and distribution of cigarettes and smokeless tobacco products to protect children and adolescents: proposed rule. *Fed Register* 1995; **60**: 41314–787.

33   Kessler DA. Statement on nicotine-containing cigarettes. *Tob Control* 1994; **3**: 148–58.

34   Food and Drug Administration. Regulations restricting the sale and distribution of cigarettes and smokeless tobacco to protect children and adolescents: final rule. *Fed Register* 1996; **61**: 44396–5318.

35   Kessler DA. The control and manipulation of nicotine in cigarettes. *Tob Control* 1994; **3**: 362–9.

36   Page JA. Federal regulation of tobacco products and products that treat tobacco dependence: are the playing fields level? *Food Drug Law J* 1998; **53**(Suppl): 11–42.

37   British Columbia Ministry of Health. *Reports on cigarette additives and ingredients and smoke constituents*. Vancouver, BC: Ministry of Health, 1998.

38   Henningfield JE, Benowitz NL, Slade J, Houston TP, *et al*. Reducing the addictiveness of cigarettes. *Tob Control* 1998; **7**: 281–93.

39   Bates C, McNeill A, Jarvis M, Gray N. The future of tobacco product regulation and labelling in Europe: implications for the forthcoming European Union directive. *Tob Control* 1999; **8**: 225–34.

40   Buck D, Godfrey C, Raw M, Sutton M. *Tobacco and jobs: the impact of reducing consumption on employment in the UK*. York: Society for the Study of Addiction, Centre for Health Economics, 1995.

41   Prabhat J, Chaloupka FJ (eds). *Curbing the epidemic: governments and the economics of tobacco control*. Washington: World Bank, 1999.

42   Joossens L, Raw M. Smuggling and cross border shopping of tobacco in Europe. *Br Med J* 1995; **310**: 1393–7.

9 | Summary and recommendations

Cigarette smoking is the single largest avoidable cause of premature death and disability in Britain, and thus presents both the greatest challenge and the greatest opportunity for all involved in improving public health. The eradication of smoking from Britain would realise massive health gains, particularly for the most disadvantaged sectors of society. The prevalence of smoking in Britain has fallen substantially since the health risks of cigarette smoking first began to be publicised, but now appears to be stabilising at approximately one in four British adults. To achieve further significant reductions in smoking prevalence it is necessary to look more radically at the causes, treatment and ultimate prevention of smoking behaviour.

The central conclusion of this report is that cigarette smoking should be understood as a manifestation of nicotine addiction, and that the extent to which smokers are addicted to nicotine is comparable with addiction to 'hard' drugs such as heroin or cocaine. This conclusion has fundamental implications for the design and implementation of public health policy on the control and prevention of cigarette smoking.

9.1 Tobacco and nicotine addiction

The unique selling point of tobacco is its nicotine content – tobacco products without nicotine are not commercially viable. Nicotine is an addictive drug, and the primary purpose of smoking tobacco is to deliver a dose of nicotine rapidly to receptors in the brain. This generates a pleasurable sensation for the smoker which, with repeated experience, rapidly consolidates into physiological and psychological addiction reinforced by pronounced withdrawal symptoms.

The presence of nicotine is necessary, but not sufficient, for the nicotine to have a powerful psychoactive impact. To achieve the latter, nicotine must also be delivered rapidly to the brain. Tobacco smoke inhalation is the most highly optimised vehicle for nicotine administration because absorption through the lungs delivers nicotine to the brain rapidly and intensively. The potency of the nicotine effect is created by the speed of delivery, not just by the total nicotine delivered. The speed of nicotine delivery is a fundamental difference between cigarettes and nicotine replacement therapy (NRT) products which deliver nicotine at lower and slower subaddictive rates. For this reason, nicotine delivered through tobacco smoke should be regarded as a powerfully addictive drug, and smoking as the means of nicotine self-administration. The risk of addiction to NRT products is very low, but they are effective in attenuating cravings and withdrawal from tobacco-delivered nicotine dependence.

In its usual dose range, nicotine use does not cause intoxication or intense euphoria, but does have a complex physiological impact which creates dependency reinforced by withdrawal. The fact that nicotine does not intoxicate does not make it less addicting, but may explain why medical bodies and governments have not generally recognised tobacco use as a form of drug addiction or dependence. It is far from clear that benefits attributed to nicotine use such as stress relief, improved mood and enhanced cognitive performance are real. Many perceived benefits are actually attributable to the relief of nicotine withdrawal symptoms.

Although nicotine in the form of tobacco is a legal drug, it should not be regarded as pharmacologically benign. The classification of drugs as 'legal', 'soft' or 'hard' reflects public perceptions and current law enforcement practice, rather than constituting a useful pharmacological classification. In terms of addictiveness, nicotine delivered in tobacco smoke is a 'hard' drug on a par with heroin and cocaine. The status of nicotine as a seemingly innocuous legal drug, and attempts for many years by the tobacco industry to equate addiction to nicotine with addiction to substances such as coffee, colas or chocolate, have distracted attention from the highly addictive nature of nicotine in cigarettes.

9.2 Consequences of nicotine addiction

The principal adverse consequences of nicotine addiction are the morbidity and mortality caused by active and passive smoking. Nicotine addiction is the primary reason why smokers find it difficult to give up smoking. Most people begin smoking and become addicted to nicotine as teenagers. This addiction may then cause tobacco use to continue long after an informed adult choice has been made to stop smoking on the grounds of a change in attitude to health, changed circumstances such as starting work in a smoke-free office, starting a family or other reasons. This characteristic of tobacco use – an attenuation of free choice initiated in childhood – is a central plank of the case for government intervention to control tobacco use through measures such as advertising bans, tax increases, anti-smoking communications and cessation support, and to regulate the availability and safety of nicotine products.

Nicotine addiction is closely linked to socio-economic disadvantage. Smoking prevalence is higher and nicotine use heavier among poorer smokers. The socio-economic gradient in smoking behaviour accounts for about two-thirds of the excess premature mortality associated with deprivation. Nicotine addiction is therefore responsible for significant health inequalities.

The addictive properties of nicotine also mean that simplistic machine measurements of tar and nicotine yields from cigarettes do not reflect real tar and nicotine exposure to smokers. Smokers adjust the way they smoke in order to self-administer a satisfactory dose of nicotine – a process known as 'compensation'. In response to reduced nicotine concentration in smoke, a smoker can adjust nicotine intake back to a satisfactory level by smoking more intensely, holding smoke in the lungs for longer, smoking more of the cigarette, or by blocking ventilation holes in the filter. Cigarette testing machines do not adjust their inhalation profile in response to changes in nicotine. This criticism of cigarette testing is not a minor point; it completely undermines the approach currently used both for regulation of tar and for consumer labelling.

The phenomenon of nicotine 'compensation' has profound implications for the regulation, labelling and branding of cigarettes. The strategy of reducing nominal tar yields has been widely and genuinely assumed by governments and the European Commission to deliver reduced harm to smokers, since it is these other products of tobacco combustion, rather than nicotine itself, that account for most of the harm caused by smoking. The improved understanding of nicotine-seeking behaviour now suggests that this assumption cannot be sustained. The unfortunate truth is that cigarettes labelled as 'low tar' do not necessarily deliver less tar to the smoker. As a result, the official labelling of cigarettes with tar yields expressed as milligrams of tar per cigarette can mislead consumers, and in the case of 'low tar' cigarettes greatly understate the health risks compared to higher tar cigarettes. In the long term, this practice may be more harmful because it may help to perpetuate smoking in people who would otherwise give up completely for health reasons. The problem is further compounded by the use of branding terms such as 'light', 'mild', and 'ultra'. Though based on the government-sanctioned tar yields, such branding makes an implied health claim for low tar cigarettes that cannot be justified in practice.

9.3 Treatment of nicotine addiction

Over two-thirds of smokers say they would like to quit and about one-third try to quit in any year, yet only about 2% succeed. Many smokers will make repeated attempts, with a period of abstinence followed by relapse.

There are two main complementary forms of intervention to assist smoking cessation:

1   Motivation, support and advice.
2   Treatment products such as NRT.

The first is designed to increase smokers' commitment to stopping, the second to help attenuate cravings and withdrawal. For both, there are approaches with proven efficacy and attractive cost-effectiveness. The benefits of smoking cessation are substantial: immediate improved health, longer life, improved welfare and finances, and reduced passive smoking exposure to family, friends and working colleagues. At present, the NHS deploys a major part of its resources on the treatment of smoking related illness, but fails to provide widely available and accessible smoking cessation or prevention services.

9.4 Regulation of tobacco products

Tobacco products have enjoyed an unprecedented degree of freedom from the safety regulations that apply to virtually all other food or drug products available in Britain. Attempts to introduce regulations on the production, safety, promotion and almost all other aspects of tobacco products have been based predominantly on voluntary agreements with the tobacco industry, and have generally failed to deliver convincing public health benefits. All forms of NRT are required, quite properly, to meet the same standards of safety and product information as any other drug product, yet nicotine delivery products based on tobacco are largely exempt from such controls. It is now time to impose appropriate, effective and consistent safety regulations on all tobacco products in Britain.

9.5 Recommendations

We advocate a broad approach to tobacco control that deploys all the available policy levers consistent with upholding civil liberties and ensuring cost-effectiveness. In 1998, the UK government published an initial strategy to address tobacco use in the UK in the White Paper, *Smoking kills*,[1] and we welcome and support all the practical initiatives proposed in that document. We recommend that this approach is augmented as follows:

1 In all areas of policy making and management of public health, nicotine delivered rapidly to the brain in tobacco smoke should be recognised as a powerfully addicting drug on a par with heroin and cocaine, and tobacco products should be recognised as nicotine delivery systems.

2 Tobacco product regulation is greatly complicated by the influence of nicotine on smoking behaviour. Current approaches to characterising the tar and nicotine yield of tobacco products are simplistic and misleading to consumers and regulators alike, and should be abandoned. This approach should be replaced with measurements and metrics that properly reflect the relative harm caused by different tobacco products, and by measures to ensure that this information is appropriately provided to consumers.

3 Harm-reduction strategies based on naïve interpretation of tar and nicotine yield measurements should be discontinued. In practical terms, this means abandoning the strategy of seeking lower nominal tar yields and instead finding approaches that genuinely reduce harm to nicotine users. Branding such as 'light', 'mild' and other words or imagery that imply a reduced health risk attributable to low tar or nicotine measurements should be banned unless and until convincing evidence of reduced health risks is forthcoming.

4 The phenomenon of nicotine dependence is heavily entrenched in society. It is obviously desirable to reduce both nicotine dependence and the terrible harm caused by nicotine delivery through tobacco smoke, but it may be necessary to accept, albeit reluctantly, the intractability of widespread nicotine dependence in the short to medium term. In this case, product developments that enable nicotine users to take nicotine with less harm to their health should be encouraged, not rejected.

5 Warning labels on tobacco products should reflect their addictive nature.

6 Consideration should be given to the use of addiction messages in communication strategies aimed at young people, for whom a loss of control may be of more concern than a risk of cancer or heart disease much later in life.

7 Raising the price of tobacco products is well recognised to be an effective incentive to quit, and is thus an important preventive measure. However, taxation of addictive drugs also has an ethical dimension in terms of the financial deprivation that it can cause. It is therefore essential that government does everything possible to facilitate smoking cessation by providing universal access to evidence-based smoking cessation services.

8 Health communication programmes and smoking cessation services should be directed with particular emphasis towards those sectors of society in which smoking is most prevalent. These currently comprise the more disadvantaged sectors of British society. This orientation is essential if the pronounced health inequalities attributable to tobacco use are to be tackled.

9 Smoking cessation services should be made universally available at all levels in the NHS, and all doctors need to recognise nicotine addiction as a major medical priority. Primary care practitioners should regard it as a core function to treat nicotine addiction, just as they would wish to address alcohol dependence or illicit drug addiction. Hospitals and all other health service providers should be required to provide appropriate cessation support to all smokers.

10 NRT is a highly effective and cost-effective smoking cessation treatment, and will play a central role in tackling nicotine addiction. NRT should be available to all smokers through reimbursable NHS prescriptions, and also be widely available and affordable for general sale. NRT products are an essential and economic option for core NHS expenditure.

11 Smoking cessation services must adapt and be adapted, as appropriate, to incorporate new therapies and interventions that are shown to be effective and cost-effective.

12 The addicted condition of smokers in the workplace must be handled with sensitivity. Employers should be encouraged to provide support for smoking cessation, and to allow time for smokers to adjust to new conditions. It is also important for employers to minimise as far as they are able the risk of relapse among smokers attempting to overcome their nicotine addiction.

13 Cigarettes are nicotine delivery products, and should be subject to the same regulatory controls as any other drug delivery device.

14 Tobacco products in Britain should therefore be regulated either by the Medicines Control Agency or by a nicotine regulatory authority similar in concept to the Food Standards Agency.

15 We recommend that an independent expert committee should be established to examine the institutional options for nicotine regulation, and to report to the Secretary of State for Health on the appropriate future regulation of nicotine products and the management and prevention of nicotine addiction in Britain.

Reference

1   Department of Health. *Smoking kills. A White Paper on tobacco*. London: The Stationery Office, 1998.